IGCSE Mathematics

module 5

University of Cambridge Local Examinations Syndicate

Reviewed by **John Pitts**, Principal Examiner and Moderator for HIGCSE Mathematics

Edited by **Carin Abramovitz**

PUBLISHED BY THE PRESS SYNDICATE OF THE UNIVERSITY OF CAMBRIDGE
The Pitt Building, Trumpington Street, Cambridge CB2 1RP, United Kingdom

CAMBRIDGE UNIVERSITY PRESS
The Edinburgh Building, Cambridge CB2 2RU, UK http://www.cup.cam.ac.uk
40 West 20th Street, New York, NY 10011-4211, USA http://www.cup.org
10 Stamford Road, Oakleigh, Melbourne 3166, Australia
Dock House, Victoria and Alfred Waterfront, Cape Town 8001, South Africa

© University of Cambridge Local Examinations Syndicate 1998

This book is in copyright. Subject to statutory exception
and to the provisions of relevant collective licensing agreements,
no reproduction of any part may take place without
the written permission of Cambridge University Press.

First published 1998
Fourth printing 2002

Printed by Creda Communications, Cape Town

Typeface New Century Schoolbook 11.5/14 pt

A catalogue record for this book is available from the British Library

ISBN 0 521 62515 7 paperback

Acknowledgements
We would like to acknowledge the contribution made to these materials by the writers and editors of the Namibian College of Open Learning (NAMCOL).

Illustrations by André Plant.

Contents

Introduction iv

Unit 1 **Perimeter and Area** 1
- A Perimeter and area of polygons 1
- B Circumference and area of circles 16
- C Extending the use of the formulae 23
- D Arcs and sectors of circles 37

Unit 2 **Surface Area and Volume** 45
- A Surface area of common solids 45
- B Volume of common solids 51
- C Extending the use of the formulae 61
- D Surface area and volume of pyramids, cones and spheres 68

Unit 3 **Right-angled Triangles and Trigonometry** 81
- A Pythagoras's theorem 81
- B Trigonometry - the tangent ratio 94
- C The sine and cosine ratios 106
- D Using trigonometry to solve problems 115

Unit 4 **Tigonometry Extended** 125
- A Trigonometry for any triangle 125
- B The sine and cosine functions 140
- C Further applications of trigonometry 143

Solutions 153

Index 173

Introduction

Welcome to Module 5 of IGCSE Mathematics! This is the **fifth module** in a course of six modules designed to help you prepare for the International General Certificate of Secondary Education (IGCSE) Mathematics examinations. Before starting this module, you should have completed Module 4. If you are studying through a distance-education college, you should also have completed the **end-of-module assignment** for Module 4. The diagram below shows how this module fits into the IGCSE Mathematics course as a whole.

Module 1	Module 2	Module 3	Module 4	Module 5	Module 6
Assignment 1	Assignment 2	Assignment 3	Assignment 4	Assignment 5	Assignment 6

Like the previous module, this module should help you develop your mathematical knowledge and skills in particular areas. If you need help while you are studying this module, contact a **tutor** at your college or school. If you need more information on writing the examination, planning your studies, or how to use the different features of the modules, refer back to the **Introduction** at the beginning of Module 1.

Some study tips for Maths

- As you work through the course, it is very important that you use a **pen or pencil and exercise book**, and *work through the examples yourself* in your exercise book as you go along. Maths is not just about reading, but also about doing and understanding!
- Do feel free to write in pencil in this book – fill in steps that are left out and make your own notes in the margin.
- *Don't expect to understand everything the first time you read it.* If you come across something difficult, it may help if you read on – but make sure you come back later and go over it again until you understand it.
- You will need a **calculator** for doing mathematical calculations and a **dictionary** may be useful for looking up unfamiliar words.

Remember

- In the examination you will be required to give decimal approximations correct to **three significant figures** (unless otherwise indicated), e.g. 14.2 or 1 420 000 or 0.00142.
- Angles should be given to **one decimal place**, e.g. 43.5°. Try to get into the habit of answering in this way when you do the exercises.

The **table** below may be useful for you to keep track of where you are in your studies. Tick each block as you complete the work. Try to fit in study time whenever you can – if you have half an hour free in the evening, spend that time studying. Every half hour counts! You can study a **section**, and then have a break before going on to the next section. If you find your concentration slipping, have a break and start again when your mind is fresh. Try to plan regular times in your week for study, and try to find a quiet place with a desk and a good light to work by. Good luck with this module!

IGCSE MATHEMATICS MODULE 5

Unit no.	Unit title	Unit studied	'Check your progress' completed	Revised for exam
1	Perimeter and Area			
2	Surface Area and Volume			
3	Right-angled Triangles and Trigonometry			
4	Trigonometry Extended			

Unit 1
Perimeter and Area

In this module, we shall deal with two parts of mathematics which are frequently applied to real-life situations. These are mensuration and trigonometry.

Mensuration is concerned with measuring. In everyday life, we may buy curtain material by **length**, we buy petrol by **volume**, and the volume of paint we need to cover a wall depends on the **area** of the wall.

The records show that, before 1000 BC, the Egyptians had methods for measuring lengths, areas and volumes. Their methods were not always accurate. For example, to find the area of a field in the shape of a quadrilateral with sides of length a, b, c, d (in that order), they used the equivalent of the formula $\frac{(a+c)}{2} \times \frac{(b+d)}{2}$ and, for a triangle with sides a, b, c, they used the formula $\frac{(a+c)}{2} \times \frac{b}{2}$. However, in most cases, their results were close enough to the correct values. There is evidence that the Babylonians and the Chinese also had methods for measuring lengths, areas and volumes, but it was the Greeks, in about 500 BC, who developed the methods logically.

In this unit I will explain to you how to calculate the perimeters and areas of 2-dimensional shapes.

This unit is divided into four sections:

Section	Title	Time
A	Perimeter and area of polygons	3 hours
B	Circumference and area of circles	2 hours
C	Extending the use of the formulae	3 hours
D	Arcs and sectors of circles	2 hours

A Perimeter and area of polygons

Measurement of length

The length of a straight line segment is a measure of the distance from one end of the line to the other. To measure the length of a straight line, you have to compare it with some unit, that means some standard length.

In the metric system, the basic unit of length is the **metre**. This was originally taken to be a ten-millionth part of the distance from the North Pole to the equator. It is now defined as 1 650 763.73 wavelengths of the orange line in the spectrum of krypton. (You will not be expected to use this in your work!) A metre is slightly more than the length of a man's stride.

Suppose you have a rod which is one metre long and you want to measure the length of a room. You could mark off distances of a metre from one end of the room to the other.

It could happen that the length of the room is a whole number of metres.

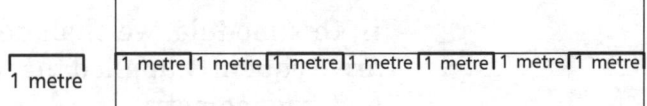

The length of this room is 7 metres.

It is more likely, however, that the length is not an exact number of metres.

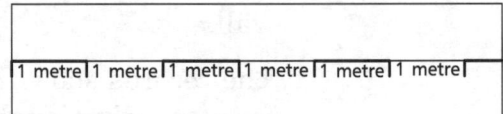

The length of this room is more than 6 metres but less than 7 metres.

In this case, you could decide to work with fractions or decimals of a metre. For example $\frac{1}{2}$ metre, $\frac{2}{3}$ metre, etc., or 0.4 metres, 0.72 metres, etc. Alternatively you could start to use a smaller unit such as a centimetre (which is one hundredth of a metre). In practice, lengths may be given using decimals (for example, 6.24 metres) or they may be given in terms of smaller units (for example, 624 centimetres) or they may be given as a mixture of units (for example, 6 metres 24 centimetres).

The units used will depend on the practical situation. If you wanted to measure the distance from Pretoria to Cape Town you would not measure it in centimetres, or even metres – these units are far too small for this purpose. On the other hand, it would be reasonable to give the height of a man in centimetres or metres.

Because there are several different units of length in common use, and mixtures of units may occur, it is necessary to know the relationship between these units.

Units of length

As you already know, the basic unit of length is the metre. Smaller units you will use are the **centimetre** and the **millimetre**. A larger unit is the **kilometre**. The relationship between these units is shown below.

1 metre = 100 centimetres	1 centimetre = 0.01 metre
1 metre = 1000 millimetres	1 millimetre = 0.001 metre
1 centimetre = 10 millimetres	1 millimetre = 0.1 centimetre
1 kilometre = 1000 metres	1 metre = 0.001 kilometre

When you are doing calculations with lengths, you'll soon get tired of writing 'metre', 'centimetre', 'millimetre' and 'kilometre' in full, you can use abbreviations, as follows:

Module 5 Unit 1

Unit	Abbreviation
millimetre	mm
centimetre	cm
metre	m
kilometre	km

Perimeter

The boundary of a plane (2-dimensional) shape is known as its perimeter. The word **perimeter** is also used to mean the **length of the boundary** of the shape.

If the shape is a polygon, calculating the length of its boundary is relatively easy because it consists of straight segments. All you have to do is to add up the lengths of the sides of the polygon.

Example 1

Calculate the perimeter of:
a) a rectangle with length 7 cm and width 5 cm
b) a rhombus with sides of length 68 mm
c) a rectangle with length 17 cm and width 8 mm

Solution

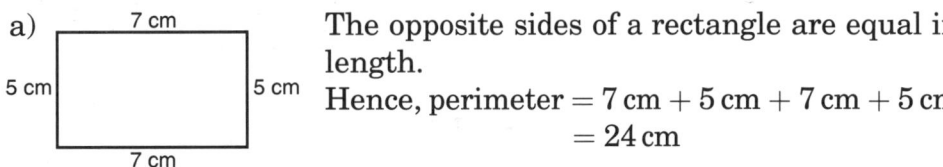

a) The opposite sides of a rectangle are equal in length.
Hence, perimeter = 7 cm + 5 cm + 7 cm + 5 cm
= 24 cm

b) A rhombus has four equal sides.
Hence, perimeter = 68 mm × 4 = 272 mm

c) The length and width are given in different units. You must express them both in centimetres or both in millimetres using 1 cm = 10 mm.
perimeter = 170 mm + 8 mm + 170 mm + 8 mm = 356 mm
or perimeter = 17 cm + 0.8 cm + 17 cm + 0.8 cm = 35.6 cm

Example 2

Calculate the perimeter of each of the following shapes. (In shape b) and in shape c), the adjacent sides are at right-angles.)

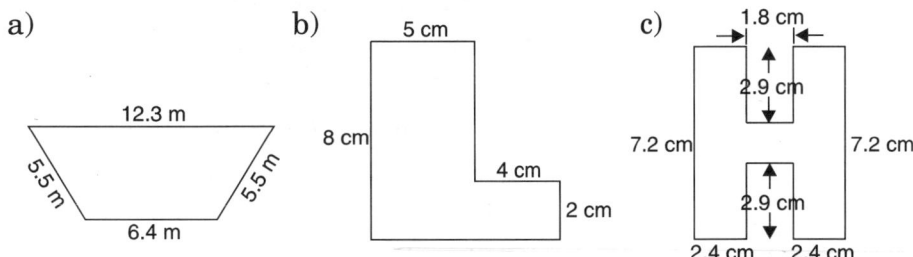

Solution

a) Perimeter = 12.3 m + 5.5 m + 6.4 m + 5.5 m = 29.7 m

b) The base of the shape = 5 cm + 4 cm = 9 cm
The unmarked vertical line = 8 cm − 2 cm = 6 cm
The perimeter of the shape
= 9 cm + 2 cm + 4 cm + 6 cm + 5 cm + 8 cm = 34 cm

Alternative method

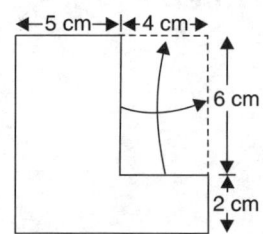

Moving the middle horizontal line and the middle vertical line, as shown in this diagram, does not alter the perimeter but it changes the shape into a rectangle with length 9 cm and height 8 cm.
Perimeter = (9 + 8 + 9 + 8) cm = 34 cm.

c) There are 2 vertical lines of length 7.2 cm and
4 vertical lines of length 2.9 cm.
There are 4 horizontal lines of length 2.4 cm and
2 horizontal lines of length 1.8 cm.

Perimeter = ((7.2 × 2) + (2.9 × 4) + (2.4 × 4) + (1.8 × 2)) cm
= (14.4 + 11.6 + 9.6 + 3.6) cm
= 39.2 cm

Example 3

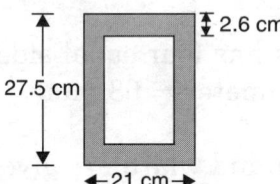

The rectangular frame for a picture is 21 cm wide and 27.5 cm high. The border is 2.6 cm wide.

Calculate the perimeter of the picture.

Solution

Width of the picture = 21 cm − (2.6 + 2.6) cm = 15.8 cm
Height of the picture = 27.5 cm − (2.6 + 2.6) cm = 22.3 cm
Perimeter of the picture = 15.8 cm + 22.3 cm + 15.8 cm + 22.3 cm
= 76.2 cm

Example 4

A rectangle has a length of L units and breadth of B units.
Its perimeter is P units.

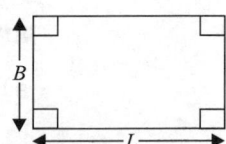

Find a formula for P in terms of L and B.

Solution

$P = L + B + L + B$
$P = 2L + 2B$

If the right-hand side is factorised, the formula becomes
$P = 2(L + B)$.

Example 5

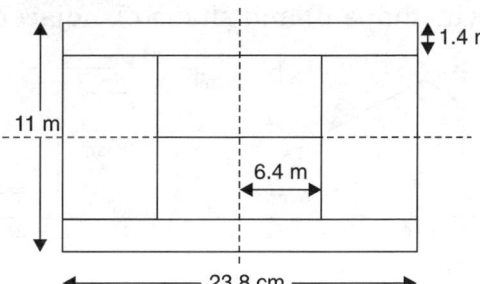

The diagram represents a tennis court. It is symmetrical about each of the dashed lines. The solid lines represent the markings on the court.

Calculate the total length of these markings.

Solution

Across the page, there are 4 lines of length 23.8 m and
1 line of length 12.8 m.
Up the page, there are 2 lines of length 11 m and
2 lines of length (11 − 1.4 − 1.4) m,
that is 8.2 m

Total length of the line = ((23.8 × 4) + 12.8 + (11 × 2) + (8.2 × 2)) m
= (95.2 + 12.8 + 22 + 16.4) m
= 146.4 m

Test your understanding of this work by answering the following questions. Don't forget to state the units in each of your answers.

EXERCISE 1

1. Calculate the perimeter of:
   a) a square with sides of length 48 mm
   b) a rectangle with length 9 cm and breadth 6 cm
   c) an equilateral triangle with sides of length 5.2 cm
   d) a quadrilateral with sides of length 3 m, 95 cm, 2.7 m and 85 cm

2. The shapes below are drawn on a grid of 1 cm squares. Calculate the perimeter of each shape.

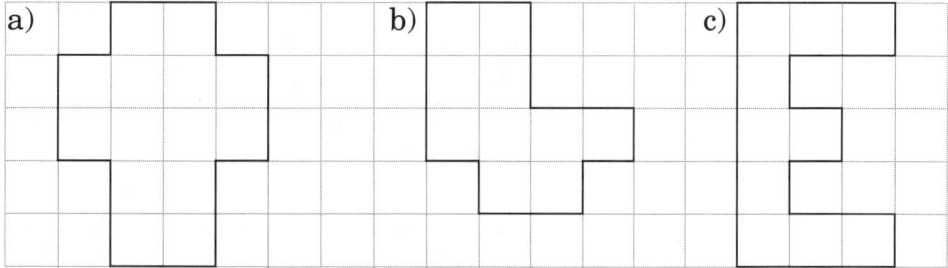

3. Calculate the perimeter of each of the following shapes.
   (In shape b) and shape c), adjacent sides are at right-angles.)

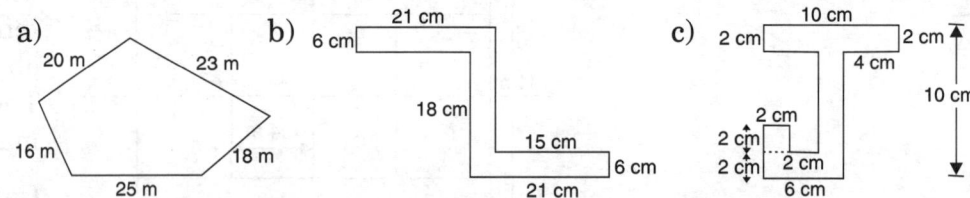

4. a) An equilateral triangle has sides of length L units.
      Its perimeter is P units.
      Write down a formula for P in terms of L.
   b) An isosceles triangle has two sides of length S units and one side of length B units.
      Its perimeter is P units.
      Write down a formula for P in terms of S and B.

5.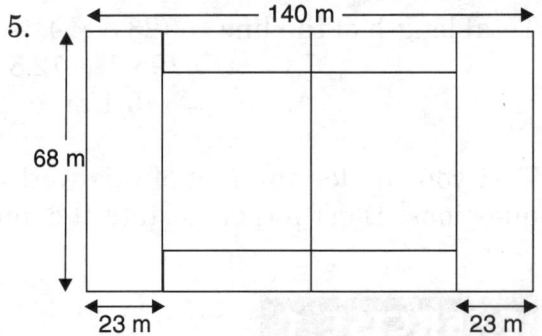

   The sketch shows the markings on a rugby field. The field is a rectangle with dimensions 140 m by 68 m.

   Calculate the total length of the markings.

6. Find, in metres, the perimeter of:
   a) a rectangle with length 85 cm and breadth 35 cm
   b) a kite with two adjacent sides of length 4.27 m and 5.83 m
   c) an isosceles trapezium with three sides of length 338 mm and one side of length 224 mm

Check your answers at the end of this module.

Measurement of area

The **area** of a plane (2-dimensional) shape is a measure of the amount of the plane it covers. To measure the area of a plane shape, you have to compare it with some unit of area. Theoretically, this could be any shape which will cover the plane without leaving any gaps.

For example, the shape ⬡ could be used, it covers the plane like this:

Or, the shape could be used, it covers the plane like this:

Taking into account that the unit shape must be as simple as possible so that we can work out fractions of it, and that we often have to find the area of a shape which contains right-angles, the shape we choose is a square.

The shape will cover the plane like this:

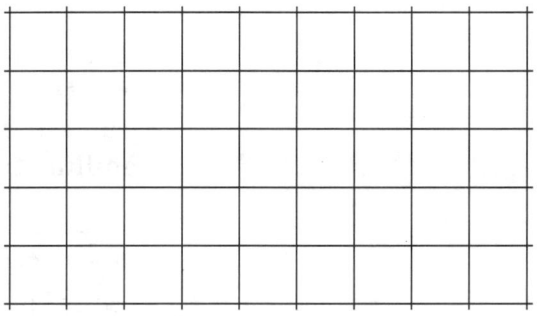

Area of a rectangle

Suppose you want to measure the area of the rectangular floor of a room. As the unit of area, you have a square card with sides of length 1 metre. This unit will be called a **square metre**.

On the floor, you mark off units of one square metre. If the length and breadth are both whole numbers of metres, you will be able to cover the floor completely.

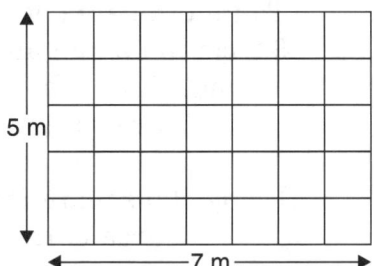

For example, if the floor is 7 m by 5 m, it can be covered by 5 rows of 7 squares, or 35 units of area. That means that the area of the floor is 35 square metres.

Can you see that, if the floor is L metres by B metres, where L and B are whole numbers, it will be covered by B rows of L squares? So the area of the floor is $L \times B$ square metres.

If the length and breadth of the floor are not both whole numbers of metres, you will have to use fractions or decimals of a square metre to find the area of the floor, or you could use a smaller unit such as a square centimetre or a square millimetre.

Here is an example using fractions:

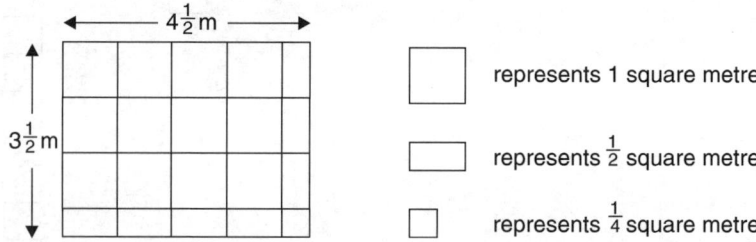

Area = 12 square metres + 7 half square metres + 1 quarter square metre
$= (12 + 3\frac{1}{2} + \frac{1}{4})$ square metres
$= 15\frac{3}{4}$ square metres

You will notice that $4\frac{1}{2} \times 3\frac{1}{2} = \frac{9}{2} \times \frac{7}{2} = \frac{63}{4} = 15\frac{3}{4}$, so the formula
area of a rectangle = length × breadth is true in this case.

If you are going to use smaller units, you need to know the relationship between the various units of area. For example, as you can see from the diagrams below, 1 square centimetre = 100 square millimetres.

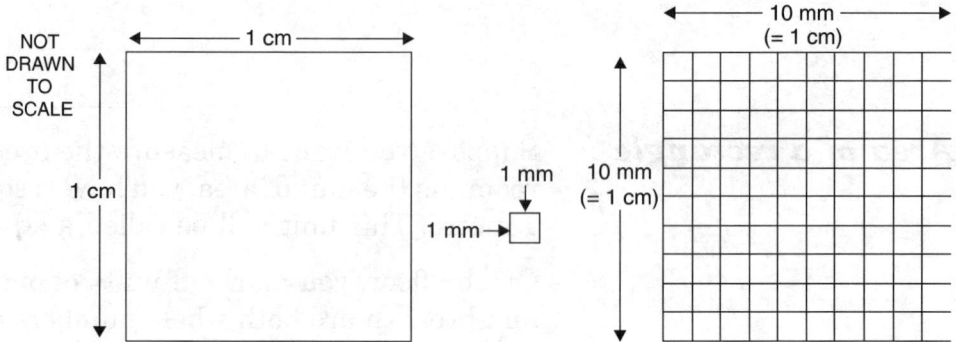

To find the area of a rectangle 9.2 cm by 6.7 cm, you first change 9.2 cm to 92 mm and 6.7 cm to 67 mm.
92 and 67 are whole numbers, so
area of the rectangle = 92 × 67 square millimetres
= 6164 square millimetres
Using 100 square millimetres = 1 square centimetre,
area of the rectangle = 61.64 square centimetres.

You will notice that 9.2 × 6.7 = 61.64, so the formula
area of rectangle = length × breadth is true in this case.

In fact, the formula is true for all values of the length and breadth, so we can state:

> area of a rectangle = length × breadth

Abbreviations

As with lengths, the units of area are very often written in an abbreviated form. For example, taking into account that a square centimetre is a square 1 cm by 1 cm, we abbreviate it to 1 cm^2.

you say 'one centimetre squared'

The abbreviations you need to know are shown below. You must be familiar with units in both forms and be able to change one form to the other, for example, square centimetres to cm² and vice versa.

Unit	Abbreviation
square millimetre	mm²
square centimetre	cm²
square metre	m²
square kilometre	km²

Units of area

The units of area are related to one another and to the units of length, as shown in this table.

Length	Area
1 cm = 10 mm	1 cm² = 10 mm × 10 mm = 100 mm²
1 m = 100 cm	1 m² = 100 cm × 100 cm = 10 000 cm²
1 km = 1000 m	1 km² = 1000 m × 1000 m = 1 000 000 m²
	1 hectare = 100 m × 100 m = 10 000 m²

Example 1

Calculate the area of a rectangle which has a length of 12.3 cm and a breadth of 8.5 cm.

Solution

Area of rectangle = length × breadth
$$= (12.3 \times 8.5) \text{ cm}^2$$
$$= 104.55 \text{ cm}^2$$

Example 2

Calculate the area of a rectangle which has a length of $4\frac{2}{3}$ m and a breadth of $4\frac{1}{2}$ m.

Solution

Area of rectangle = length × breadth
$$= (4\tfrac{2}{3} \times 4\tfrac{1}{2}) \text{ m}^2$$
$$= \tfrac{14}{3} \times \tfrac{9}{2} \text{ m}^2$$
$$= 21 \text{ m}^2$$

Example 3

Calculate the area of a rectangle which has a length of 2.4 m and a breadth of 85 cm.

Solution

The length and breadth must be expressed in the same units. You could choose to work in metres or in centimetres.

length = 2.4 m and breadth = 0.85 m
area = (2.4 × 0.85) m^2
= 2.04 m^2

or length = 240 cm and breadth = 85 cm
area = (240 × 85) cm^2
= 20 400 cm^2

Example 4

A mat has an area of 52 500 cm^2. Express this area in square metres.

Solution

1 square metre = 1 metre × 1 metre
= 100 cm × 100 cm
= 10 000 cm^2

Hence, 52 500 cm^2 = $\frac{52\ 500}{10\ 000}$ m^2
= 5.25 m^2

Here are a few questions for you to try.

EXERCISE 2

1. Calculate the area of a rectangle with:
   a) a length of 9.5 cm and a breadth of 7.4 cm
   b) a length of $3\frac{1}{2}$ m and a breadth of $2\frac{2}{5}$ m
   c) a length of 5.6 m and a breadth of 75 cm

2. Calculate the area of a square with sides of 8.6 cm.

3. Calculate the area, in square centimetres, of a rectangle with:
   a) a length of 425 mm and a breadth of 220 mm
   b) a length of 1.35 m and a breadth of 0.86 m

4. A field has an area of 0.00357 km^2. Express this area in square metres.

5. A soccer field is a rectangle 125 m by 90 m. Calculate its area in hectares.

Check your answers at the end of this module. I hope that most of your answers are correct. If not, consider whether you used the same units for the length and breadth, whether you changed units correctly, and, if you used a calculator, whether you have used it correctly.

Now that you know how to calculate the area of a rectangle, you can easily find methods for calculating the areas of other polygons.

Area of a triangle

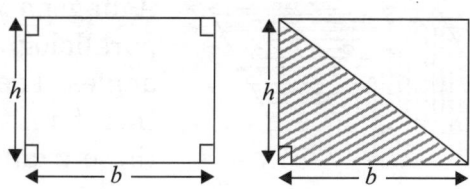

The diagonal of a rectangle cuts the rectangle into two congruent right-angled triangles. It follows that the area of each right-angled triangle is half the area of the rectangle.

In a right-angled triangle, we take the arms of the right angle to be b (for 'base') and h (for 'height').

When we have a triangle which is not right-angled, we divide it into two right-angled triangles and consider each of them as half a rectangle.

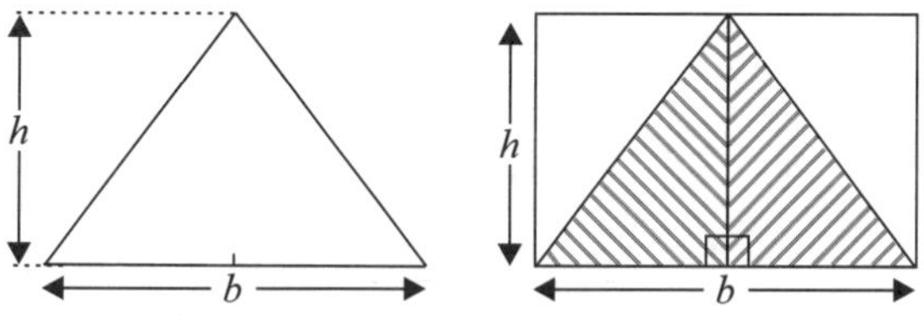

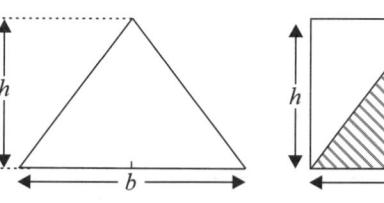

The two rectangles together make a large rectangle. The area of the whole triangle is half the area of the large rectangle. Hence, once again, we have area of triangle $= \frac{1}{2}(b \times h)$.

You will realise that the base (b) of the triangle is one of the sides and the height (h) is measured perpendicular to the base. Usually, h is *not* one of the sides (it could be if the triangle is right-angled). For this reason we usually refer to h as the **perpendicular height**.

As we can take any of the three sides of the triangle as the base, there are three different perpendicular heights. (See the diagrams below.) When you are calculating the area of a triangle, you must pair the base with the correct perpendicular height.

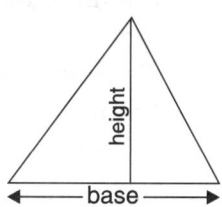 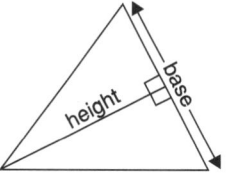 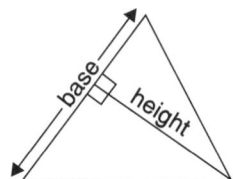

If the triangle is obtuse-angled, the 'height' line could be outside the triangle.

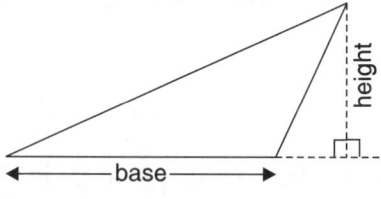

The formula for the area is the same as before (although it is a little bit more difficult to prove).

For all triangles we have the result:

> area of a triangle $= \frac{1}{2}$ (base × perpendicular height)

Any polygon can be divided up into triangles, so we can now say that we can find the area of any polygon. We will find formulae for working out the areas of certain polygons – for example, parallelograms and trapeziums.

Area of a parallelogram

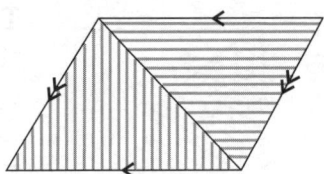

A diagonal of a parallelogram divides the parallelogram into two congruent triangles. The area of the parallelogram is, therefore, twice the area of either of these triangles.

We can now state that:

> area of a parallelogram = base × perpendicular height

You should notice that you can choose which side of the parallelogram to take as the base, but you must be careful to pair it with the correct perpendicular height.

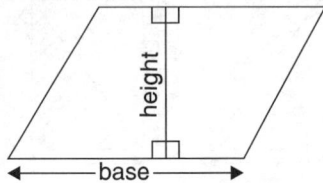 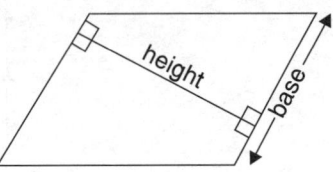

In each case, the 'perpendicular height' is the distance between a pair of opposite (parallel) sides of the parallelogram.

Area of a trapezium

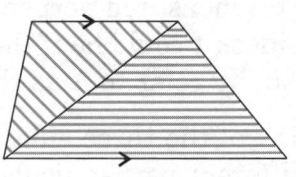

A trapezium has one pair of sides parallel. It can be divided into two triangles by a diagonal, but these triangles are not congruent.

Suppose that the lengths of the parallel sides are a and b, and that the perpendicular distance between the parallel sides is h.

Taking a and b as the bases of the two triangles, the perpendicular height of each triangle is h.

Area of the triangle with base $a = \frac{1}{2}(a \times h)$

Area of the triangle with base $b = \frac{1}{2}(b \times h)$

Adding these, we obtain area of trapezium $= \frac{1}{2}(ah) + \frac{1}{2}(bh)$
$$= \frac{1}{2}(ah + bh)$$
$$= \frac{1}{2}(a + b)h$$

We now use the fact that $\frac{1}{2}(a + b)$ is the average of a and b, and obtain the formula:

> area of a trapezium = the average of the parallel sides × the perpendicular distance between them

Example 1

1. Calculate the area of each of the following shapes.

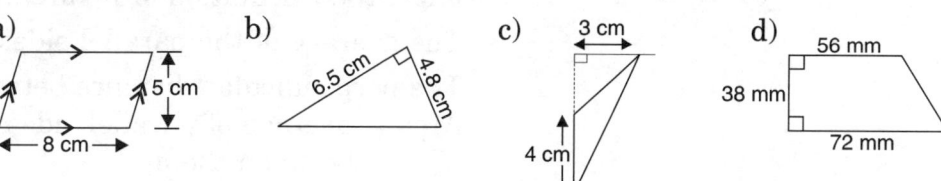

Solution

a) The shape is a parallelogram with base 8 cm and perpendicular height 5 cm.
Area = base × perpendicular height = (8 × 5) cm² = 40 cm²

b) This is a right-angled triangle.
Take 6.5 cm as the base. Then the perpendicular height is 4.8 cm.
Area = $\frac{1}{2}$(base × perpendicular height)
= $\frac{1}{2}$(6.5 × 4.8) cm² = 15.6 cm²

c) The shape is a triangle. Take 4 cm as the base. Then the perpendicular height is 3 cm.
Area = $\frac{1}{2}$(base × perpendicular height) = $\frac{1}{2}$(4 × 3) cm² = 6 cm²

d) The co-interior angles add up to 180° so the top and bottom sides of the shape are parallel. The shape is, therefore, a trapezium.
The average of the parallel sides = $\frac{1}{2}$(56 + 72) mm = 64 mm
Area = average of parallel sides × perpendicular distance between them
= (64 × 38) mm² = 2432 mm²

Example 2

The shapes below are drawn on a grid of $\frac{1}{2}$ cm squares.
Calculate the area of each shape.

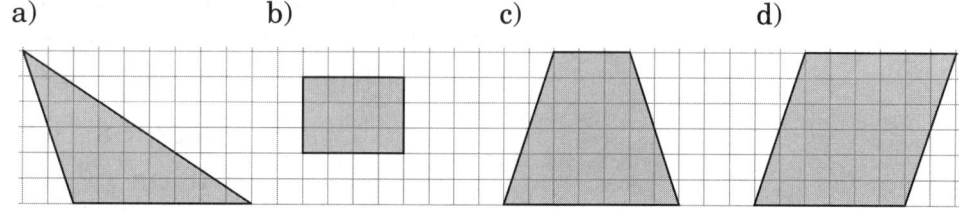

Solution

a) The horizontal side of the triangle = 7 small squares = 3.5 cm
The perpendicular height = 6 small squares = 3 cm
Area of triangle = $\frac{1}{2}$(base × perpendicular height)
= $\frac{1}{2}$(3.5 × 3) cm²
= 5.25 cm²

b) The shape is a rectangle with length 2 cm and breadth 1.5 cm
Area of rectangle = length × breadth
= (2 × 1.5) cm²
= 3 cm²

c) The shape is a trapezium. (It is an **isosceles** trapezium because the two non-parallel sides are the same length. But this does not affect the calculation of its area.)

The average of the parallel sides $= \frac{1}{2}(3.5 + 1.5)$ cm $= 2.5$ cm

The perpendicular distance between the parallel sides $= 3$ cm

Area = average of parallel sides × perpendicular distance between them
$= (2.5 \times 3)$ cm^2
$= 7.5$ cm^2

d) The horizontal base of this parallelogram $= 3$ cm
and the corresponding perpendicular height $= 3$ cm
Area = base × perpendicular height
$= (3 \times 3)$ cm^2
$= 9$ cm^2

Example 3

Calculate the area of each of the following shapes.

a) 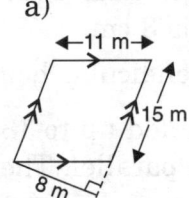 b) 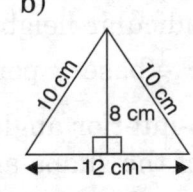 c) 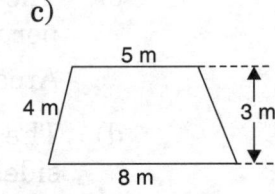 d)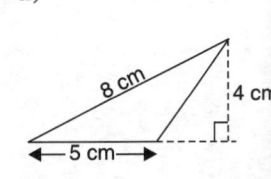

Solution

a) Taking 15 m as the base of the parallelogram, the perpendicular height is 8 m. (Do not take 11 m as the base because the corresponding perpendicular height is not known.)
Area of parallelogram = base × perpendicular height
$= (15 \times 8)$ m^2
$= 120$ m^2

b) Taking the 12 cm side as the base of the triangle, the corresponding perpendicular height is 8 cm.
Area of triangle $= \frac{1}{2}$(base × perpendicular height)
$= \frac{1}{2}(12 \times 8)$ cm^2
$= 48$ cm^2

c) The shape is a trapezium with parallel sides of 5 m and 8 m. The parallel sides are 3 m apart.
Area = average of the parallel sides × perpendicular distance between them
$= \frac{5+8}{2} \times 3$ m^2
$= (6.5 \times 3)$ m^2
$= 19.5$ m^2

d) Taking the 5 cm side as the base of the triangle, the corresponding perpendicular height is 4 cm.

Area of triangle $= \frac{1}{2}$(base × perpendicular height)

$= \frac{1}{2}(5 \times 4)$ cm^2

$= 10$ cm^2

Some of the work we have done in this module was probably familiar to you. Now try the following exercise. Don't forget to state the units for each of your answers.

EXERCISE 3

1. Calculate the area of each of the following shapes.

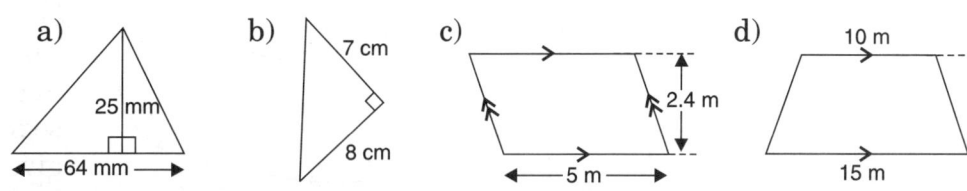

2. The shapes below are drawn on a grid of $\frac{1}{2}$ cm squares. Calculate the area of each shape.

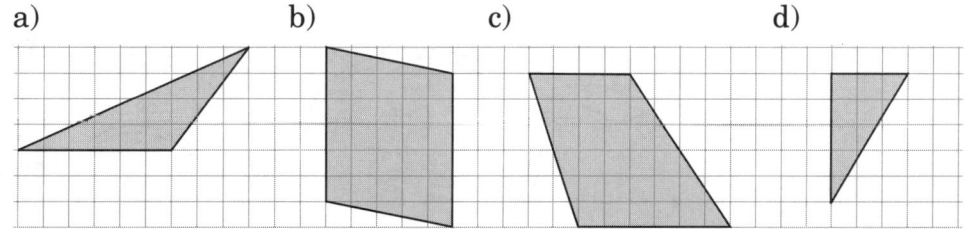

3. Calculate the area of the following shapes.

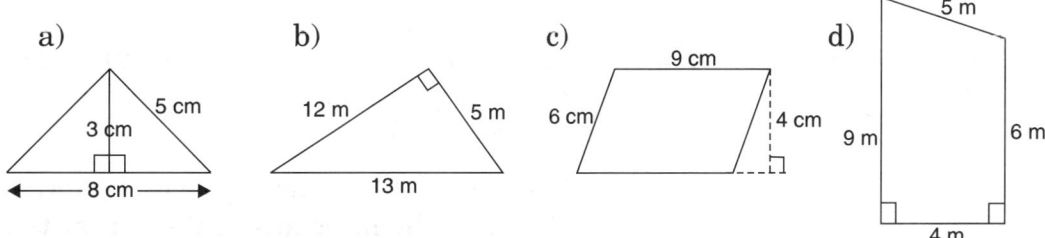

Check your answers at the end of this module before you move on to consider plane shapes which are not polygons.

B Circumference and area of circles

Perimeter of a curved shape

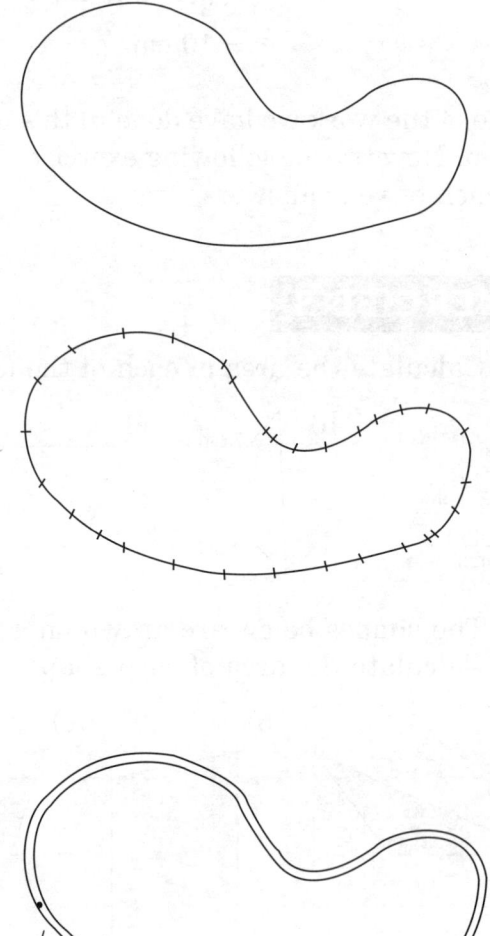

Here is a shape with a curved boundary.

How can we measure its perimeter?

One method is to divided the boundary into short sections and replace each section by a straight line segment with the same ends. The shape then becomes a polygon and we can calculate the perimeter by adding up the lengths of the line segments. This method will give a reasonably good approximation to the perimeter provided the line segments are short enough. (It will always give an *under*-estimate. Can you see why?)

Another method is to lay a piece of thin string along the boundary so that it just covers the whole of it, straighten out the string and measure its length.

Circumference of a circle

The perimeter of a circle is usually called its circumference. You can find the circumference of a circle using either of the methods mentioned above.

More than 2000 years ago, the Egyptian, Babylonian, Chinese and Greek mathematicians realised that there is a connection between the circumference and diameter of a circle. You can find this relationship for yourself by measuring the circumference and diameter of various circular objects such as a bicycle wheel, a coin, a waste paper bin, a saucer, a plate, a cocoa tin, the centre circle on a soccer field, etc.

In some cases, you could use the string method. With the waste bin and the cocoa tin, you should wrap the string round 10 times and divide the length of string used by 10; this improves the accuracy of your result. In other cases, you could roll the object along a straight line until it has made one, two, three ... complete turns. (The circumference of a wheel is equal to the distance the wheel moves forward in one turn.)

Your results could look something like this:

Object	Circumference	Diameter	$\frac{\text{Circumference}}{\text{Diameter}}$
coin	89 mm	28 mm	3.18
cocoa tin	215 mm	68 mm	3.16
cup	255 mm	81 mm	3.15
saucer	450 mm	144 mm	3.13
small plate	528 mm	168 mm	3.14
waste bin	650 mm	206 mm	3.16
large plate	772 mm	245 mm	3.15
wheel	206 cm	66 cm	3.12
centre circle (soccer field)	57.5 m	18.3 m	3.14

Note: The circumference and diameter must be measured in the same units.

You should be able to see from your results (and from those above) that the circumference of a circle is slightly more than 3 times the diameter. The Chinese, in the 12th century BC, and later the Babylonians and Jews, used 3 as the multiplier. The Greeks tried to find a more accurate multiplier and Archimedes found that it was between $3\frac{1}{7}$ and $3\frac{10}{71}$. In about AD 450, Chinese mathematicians gave $\frac{355}{113}$ as the multiplier. We now know that this is a very good approximation.

Using advanced mathematics and computers, mathematicians have worked out the value of the multiplier to thousands of significant figures. Correct to 10 significant figures, it is 3.141592654. No matter how many figures we give, the value is still an approximation. The fact is that it is impossible to find an exact fraction or decimal for it. We therefore give it a special name, **pi**, and use a special symbol, π, for it. (This is a letter in the Greek alphabet.)

We can now write the formula:

> circumference of a circle = $\pi \times$ diameter

If we use d to represent the diameter and r to represent the radius, and remember that the diameter is twice the radius, we can write:

> circumference of a circle = πd or $2\pi r$

Depending on how accurately the diameter (or radius) has been measured, and how accurate the answer has to be, we shall take π to be 3 or 3.1 or 3.14 or 3.142 or 3.1416.

When you are doing calculations involving π, you may find it convenient to use the special key for π on your calculator. This value of π is correct to more than 5 significant figures, but you must remember that, in most practical problems, the measurements are correct to at most 3 or 4 figures. When answering examination questions, you should give your answers correct to 3 significant figures unless you are told to use some other degree of accuracy.

Example 1

Calculate the circumference of a circle with:
a) a diameter of 6.25 cm
b) a diameter of 18.2 m
c) a radius of 74 mm

Solution

a) Circumference = πd = 3.142 × 6.25 cm
   = 19.6375 cm
   = 19.6 cm to 3 significant figures

b) Circumference = πd = 3.142 × 18.2 m
   = 57.1844 m
   = 57.2 m to 3 significant figures

c) Circumference = $2\pi r$ = 2 × 3.142 × 74 mm
   = 465.016 mm
   = 465 mm to 3 significant figures

Example 2

The minute hand of a church clock is 1.23 m long.
How far does the tip of the hand move in 1 hour?

Solution

The hand makes one complete revolution in an hour.
The tip moves along a circle of radius 1.23 m.
Distance moved by the tip = $2\pi r$ = 2 × 3.142 × 1.23 m
   = 7.72932 m
   = 7.73 m to 3 significant figures

Example 3

A bicycle wheel has a diameter (including the tyre) of 66 cm. Calculate how far, in metres, the wheel will go forward in 50 revolutions.

Solution

In one revolution, the wheel will go forward a distance equal to its circumference. Circumference of the wheel $= 2\pi r = 2 \times 3.142 \times 66$ cm
$= 207.372$ cm

Distance wheel goes forward in 50 revolutions $= 207.372 \times 50$ cm
$= 10368.6$ cm
$= 103.686$ m $\boxed{1\,\text{m} = 100\,\text{cm}}$
$= 104$ m
to 3 sig. figures

Note: The 'correction' to 3 significant figures is done at the very last step. It would be a mistake to correct the circumference of the wheel to 3 figures because your final answer would be less accurate.

The following questions are all concerned with the circumference of a circle. You need to decide whether you will be using the radius or the diameter, and then apply the appropriate formula, πd or $2\pi r$.

EXERCISE 4

1. Calculate the circumference of a circle with:
   a) a radius of 20 m
   b) a radius of 12.3 cm
   c) a diameter of 57 mm

2. Taking the radius of the Earth to be 6400 km, calculate the length of the equator.

3. Cotton is wound on a reel which has a diameter of 2.3 cm. There are 900 turns of cotton on the reel. What is the total length, in metres, of the cotton on the reel?

Check your answers at the end of this module.

Area of a curved shape

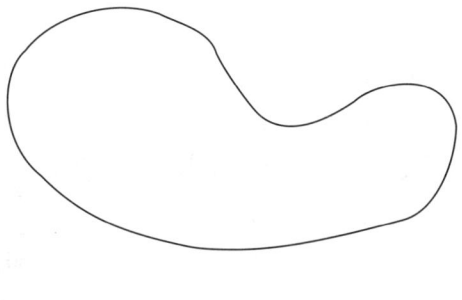

We discussed how to find the perimeter of this shape with a curved boundary. We will now consider how to find its area.

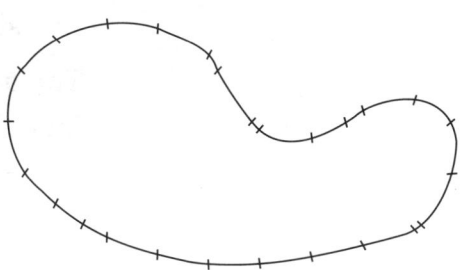

We could mark points on the boundary and join them up to form a polygon. The area of the polygon will be approximately equal to the area of the curved shape, provided the parts of the polygon outside the shape are reasonably well balanced by the parts of the shape outside the polygon.

Alternatively, we could put the curved shape on a grid of unit squares and count the squares inside the shape.

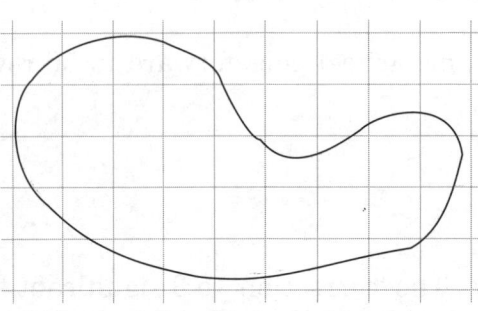

We would have to decide what to do about the squares which are not completely inside the shape.

Usually, if more than half a square is inside the shape, we count it as a whole square. If exactly half a square is inside the shape, we count it as a half square. If less than half a square is inside the shape, we do not count it.

Example 1 and solution

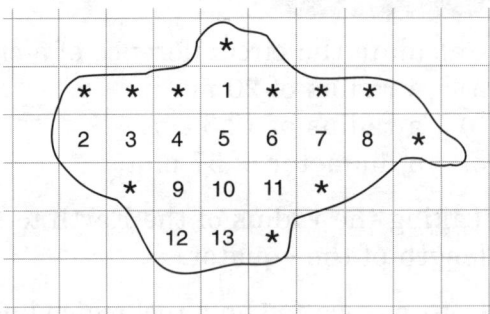

This is a map of an island on a grid of squares of side 1 km.

The 13 squares numbered 1 to 13 are completely inside the island. The 11 squares marked with an asterisk (*) are each more than half inside the island.

The area of the island is approximately $(13 + 11)$ km^2, that is 24 km^2.

Example 2 and solution

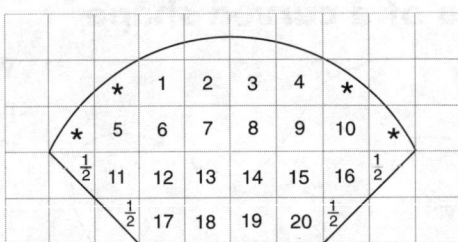

This shape is drawn on a grid of squares of side 1 cm.

The 20 squares numbered 1 to 20 are completely inside the shape. The 4 squares marked with an asterisk (*) are each more than half inside the shape. The 4 squares marked $\frac{1}{2}$ are each exactly half inside the shape.

The area of the shape is approximately $(20 + 4 + 4 \times \frac{1}{2})$ cm^2, that is 26 cm^2.

Example 3 and solution

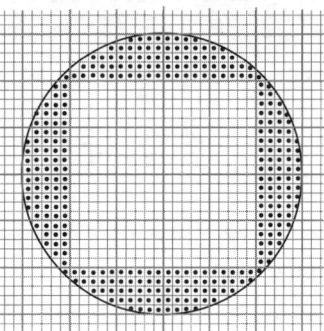

This is a circle drawn on 2 mm graph paper.

There are 400 unmarked small squares in the centre. The other small squares to be counted (because each is more than half inside the circle) are dotted. There are 292 of them.

Area of the circle = 692 small squares
1 small square = 2 mm × 2 mm
$= 4$ mm^2
Area of the circle = 692 × 4 mm^2
$= 2768$ mm^2

Since 1 cm^2 = 100 mm^2, we could write area of the circle = 27.68 cm^2, but taking into account that this is only an approximation, we will say that the area of the circle is about 28 cm^2.

Area of a circle

As you have seen, we can find the area of a circle by counting squares. This is a tedious method and the results may not be very accurate. We will look for a formula similar to the one we have for the circumference of a circle.

Consider this circle which has a radius r. We know that its circumference is $2\pi r$.

It has been divided into 8 equal parts, and 1 of these parts has been divided into 2 halves. (Each of these parts is a **sector** of the circle.)

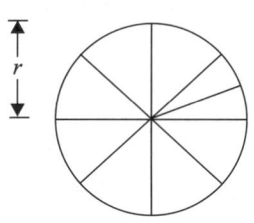

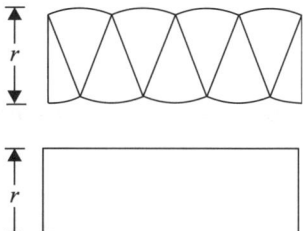

The 9 sectors have been rearranged. The total area is the same as the area of the circle but the shape is now roughly a rectangle. (If the circle had been divided into, say, 800 equal parts the shape of the rearrangement would be almost indistinguishable from a rectangle.)

The breadth of the rectangle is r (approximately). The length of the rectangle is roughly the same as the total length of the arcs at the top (or the total length of the arcs at the bottom). The 4 arcs at the top originally formed half the circumference, so their total length is $\frac{1}{2}(2\pi r)$, that is πr.

Hence, the length of the rectangle = πr (approximately).

The area of the rectangle = length × breadth
$= \pi r \times r$ (approximately)
$= \pi r^2$

From this demonstration, we would expect the area of a circle to be about πr^2. In fact, it can be proved (using some advanced mathematics) that the area of a circle is *exactly* πr^2.
You must remember this result:

$$\text{area of a circle} = \pi \times (\text{radius})^2$$

If we use r to represent the radius, this becomes:

$$\text{area of a circle} = \pi r^2$$

Note: You have to be careful when using this formula – learners often make mistakes!
You must make sure that you *use the radius* of the circle, *not the diameter*.
You must remember that it is only the radius which is squared – *do not work out (πr) squared*. (You could fall into this trap if you are using a calculator.)

Example 1

Calculate the area of a circle which has a radius of 4.75 m.

Solution

$$\begin{aligned}\text{Area of the circle} = \pi r^2 &= 3.142 \times (4.75)^2 \text{ m}^2 \\ &= 70.891375 \text{ m}^2 \\ &= 70.9 \text{ m}^2 \text{ to 3 significant figures}\end{aligned}$$

Note: If you use the π key on your calculator, you will get 70.88218425 m^2 which, corrected to 3 figures, is 70.9 m^2. Generally speaking, answers have to be given to 3 figures and you will get the same final answer using 3.142 for π as you do by using the calculator key for π.

Example 2

A car's brake disc has a diameter of 7.9 cm. What is its area?

Solution

Here you must notice that it is the *diameter* which is given, so you must first find the radius.

$$\text{radius} = \tfrac{1}{2}(7.9 \text{ cm}) = 3.95 \text{ cm}$$
$$\begin{aligned}\text{Area of disc} = \pi r^2 &= 3.142 \times (3.95)^2 \text{cm}^2 \\ &= 49.023055 \text{ cm}^2 \\ &= 49.0 \text{ cm}^2 \text{ to 3 significant figures}\end{aligned}$$

Example 3

A circular radar screen has a diameter of 38 cm. Only 90% of its area is effective. Calculate the effective area.

Solution

The *diameter* of the screen = 38cm, so the radius = 19 cm
Total area of screen = πr^2 = 3.142 × (19)² cm²
Effective area = 90% of 3.142 × (19)² cm²
 = 0.9 × 3.142 × (19)² cm²
 = 1020.8358 cm²
 = 1020 cm² to 3 significant figures

You should now be ready to tackle some questions on areas of curved shapes. Here are a few to test your understanding.

EXERCISE 5

1.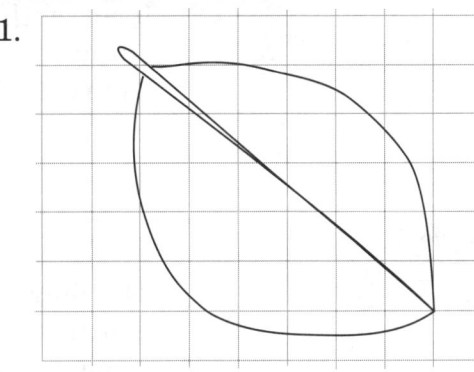

    This diagram represents the leaf of a beech tree. It is drawn on a grid of squares of side 1 cm. Estimate the area of the leaf in square centimetres.

2. Calculate the area of a circle with:
    a) a radius of 5.2 cm
    b) a diameter of 1.56 m
    c) a radius of 23 mm

3. A circus ring is a circle with diameter 15 m. Calculate the area of the ring.

4. A circular dance floor has a diameter of 10.4 m. The floor is to be sealed. One drum of sealant will cover 9 m². How many drums of sealant have to be bought?

Check your answers at the end of this module.

C Extending the use of the formulae

If a shape can be split up into rectangles, triangles, parallelograms and/or semicircles, we can find its area by adding up the areas of the various parts. (The area of a semicircle is, of course, just half the area of the circle.) Sometimes a shape can be regarded as a rectangle with pieces cut off, in which case the area of the shape can be found by subtraction.

Example I

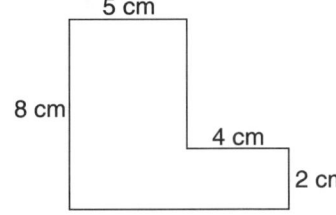

In this shape, adjacent sides are at right angles.

Find the area of the shape.

Solution

Method of addition

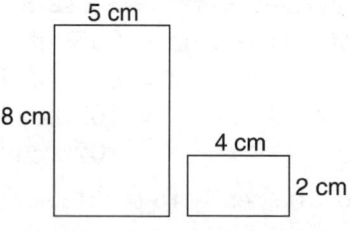

Split the shape up into two rectangles and use area = length × breadth for each of them.

Area of the shape = $((8 \times 5) + (4 \times 2))$ cm^2
= $(40 + 8)$ cm^2
= 48 cm^2

Method of subtraction

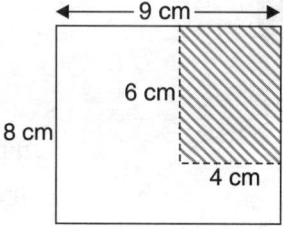

Regard the shape as a rectangle 8 cm by 9 cm, with a rectangle 6 cm by 4 cm cut off.

Area of the shape = $((8 \times 9) - (6 \times 4))$ cm^2
= $(72 - 24)$ cm^2
= 48 cm^2

Example 2

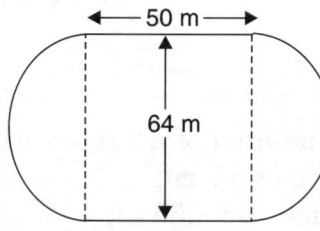

The diagram represents a sports field which consists of a rectangle and two semicircles.
a) Calculate the area of the field.
b) Calculate the length of the boundary of the field.

Solution

a) The area of the rectangle = length × breadth
= (64×50) m^2 = 3200 m^2
The two semicircles put together make a circle of radius 32 m.
Total area of the semicircles = $\pi r^2 = 3.124 \times (32)^2$ m^2
= 3217.408 m^2
Area of the sports field = $(3200 + 3217.408)$ m^2
= 6417.408 m^2
= 6420 m^2 to 3 significant figures

b) The boundary of the field consists of two straight lines, each of length 50 m, and two semicircles.
The two semicircles together make a circle of diameter 64 m.
The total length of the boundary = $(50 + 50 + (3.142 \times 64))$ m
= $(100 + 201.088)$ m
= 301.088 m
= 301 m to 3 significant figures

Example 3

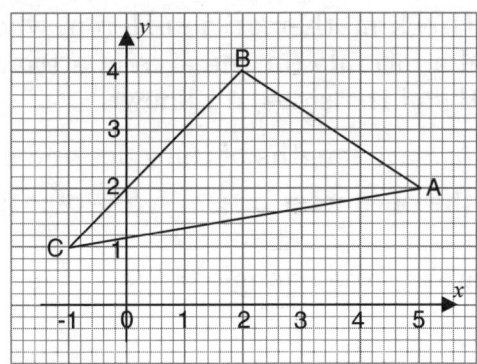

The diagram shows a triangle ABC with vertices A (5, 2), B (2, 4) and C (−1, 1).
Calculate exactly the area of the triangle.

Solution

It is difficult to calculate the area of this triangle exactly using area = $\frac{1}{2}$(base × perpendicular height). It is much easier to use the method of subtraction.

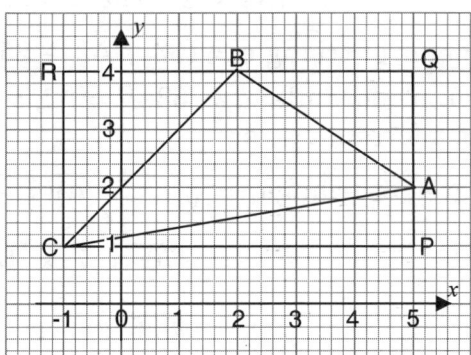

The triangle is 'boxed' in a rectangle PQRC as shown in this diagram, and then the right-angled triangles ACP, ABQ and BCR are 'taken away'.

$$\begin{aligned}
\text{Area of rectangle PQRC} &= \text{length} \times \text{breadth} \\
&= 6 \times 3 \text{ units}^2 \\
&= 18 \text{ units}^2 \\
\text{Area of triangle ACP} &= \tfrac{1}{2}(6 \times 1) \text{ units}^2 \\
&= 3 \text{ units}^2 \\
\text{Area of triangle ABQ} &= \tfrac{1}{2}(3 \times 2) \text{ units}^2 \\
&= 3 \text{ units}^2 \\
\text{Area of triangle BCR} &= \tfrac{1}{2}(3 \times 3) \text{ units}^2 \\
&= 4.5 \text{ units}^2
\end{aligned}$$

Hence, area of triangle ABC = (18 − 3 − 3 − 4.5) units^2
= 7.5 units^2

> I have worked in 'units', not in centimetres. In this way, the final answer is independent of the scales used on the coordinate axes.

Example 4

The Greek mathematician Heron, who lived in the first century AD, found a formula for the area of a triangle which is useful when the lengths of the sides are known.

If a, b, c are lengths of the sides and s is the semi-perimeter (that is half the perimeter or $\frac{1}{2}(a + b + c)$), then

the area of the triangle $= \sqrt{s(s - a)(s - b)(s - c)}$.

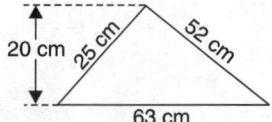

Check that Heron's formula gives the correct answer for the area of this triangle.

Solution

Area triangle $= \frac{1}{2}$(base × perpendicular height)

$= \frac{1}{2}(63 \times 20)$ cm^2 = 630 cm^2

Perimeter of triangle $= (63 + 52 + 25)$ cm $= 140$ cm

In Heron's formula, $a = 63$, $b = 52$, $c = 25$ and $s = \frac{140}{2} = 70$

Heron's formula gives

area of triangle $= \sqrt{s(s - a)(s - b)(s - c)}$

$= \sqrt{70 \times 7 \times 18 \times 45}$ cm^2

$= \sqrt{396\,900}$ cm^2

$= 630$ cm^2

This is the same as the result obtained by the usual formula.

Example 5

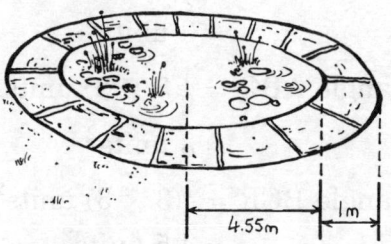

A circular pond has a radius of 4.55 m.

The pond is surrounded by a concrete path 1 m wide.

Calculate the area of the surface of the path correct to 2 significant figures.

Solution

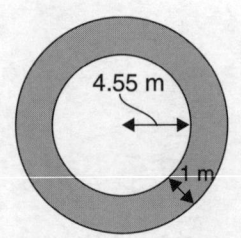

The outside edge of the path is a circle with radius 5.55 m.

The path can be regarded as a circle of radius 5.55 m with a circle of radius 4.55 m taken away.

> Knowledge of algebraic factorisation can be quite useful in mensuration questions. However, in this question, you could work out $\pi(5.55)^2$ and $\pi(4.55)^2$ separately and obtain area of path $= (96.781455 - 65.047255)$ m^2 $= 31.7342$ m^2

Area of the path $= (\pi(5.55)^2 - \pi(4.55)^2)$ m^2
$= \pi((5.55)^2 - (4.55)^2)$ m^2 [taking out the common factor π]
$= \pi(5.55 - 4.55)(5.55 + 4.55)$ m^2
$= (3.142 \times 1 \times 10.10)$ m^2 [using $a^2 - b^2 = (a - b)(a + b)$]
$= 31.7342$ m^2
$= 32$ m^2 to 2 significant figures

Notice that the answer is given to 2 significant figures and not the usual 3 significant figures. This follows the instruction in the question.

You may have found this work more difficult to follow than the previous work on areas. You should try some questions for yourself before you do any more new work. Remember to show your working – in the IGCSE examinations, you may be given marks for showing that you know the correct method even if your final answer is wrong.

EXERCISE 6

1. In each of the diagrams below, the dimensions are in centimetres.
   Find the area of each shape.

   a) 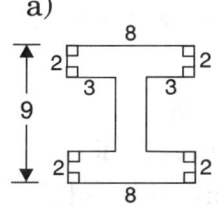   b) 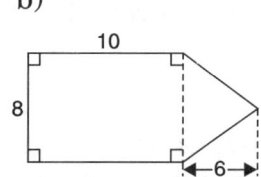   c) 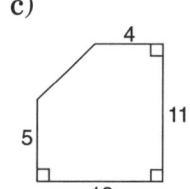   d)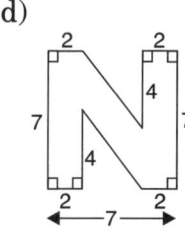

2.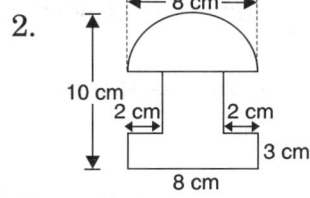

   Find the area of this shape.
   Give your answer correct to 2 significant figures.

3.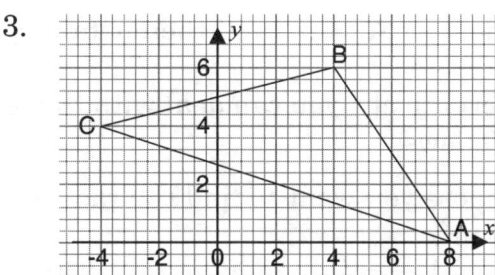

   In the diagram, point A has coordinates (8, 0) and point B has coordinates (4, 6).
   a) Write down the coordinates of point C.
   b) Calculate the exact area, in square units, of triangle ABC.

4.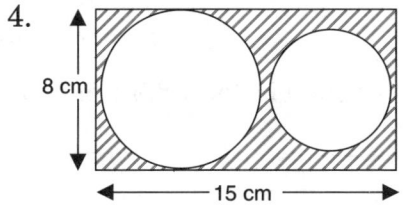

   Two circular discs, of radii 4 cm and 3 cm, are cut from a rectangular piece of card measuring 15 cm by 8 cm, as shown in the diagram.

   Calculate the area of the remaining card.

5.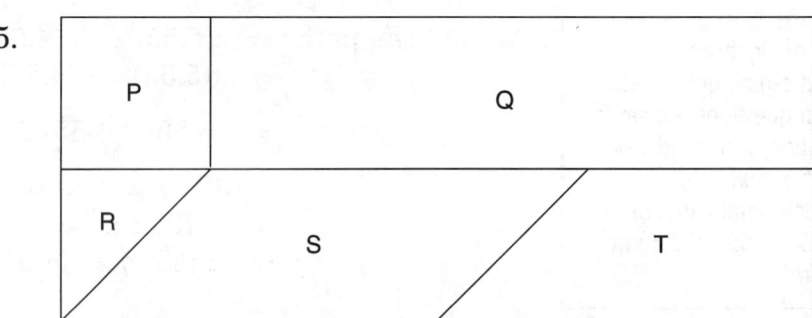

The diagram is a scale drawing of a rectangular wall painting. The scale is 1 cm to 1 m, and each of the lengths needed is a whole number of centimetres.
a) Take suitable measurements and then calculate the areas on the wall of shapes P, Q, R, S and T.
b) Areas P and T are painted red, area Q blue, area R green and area S yellow.
 (i) What fraction of the whole painting is blue? Give your answer in its lowest terms.
 (ii) What percentage of the whole painting is red?

Check your answers at the end of this module.

Changing the subject

The formula $A = \pi r^2$ is useful when you know the radius (r) of a circle and you want to find the area (A). But suppose your problem is the other way round — you know the area and you want to find the radius.

In this case, you will have to change the subject of the formula from A to r. Changing the subject of a formula is a topic which you studied in Module 2. Dividing both sides of $A = \pi r^2$ by π, we get $\frac{A}{\pi} = r^2$.

Then, taking the square root of both sides, we get $\sqrt{\frac{A}{\pi}} = r$.

Thus, the formula for finding the radius (r) of a circle when you know its area (A) is $r = \sqrt{\frac{A}{\pi}}$.

There is no need for you to remember this formula — you should remember $A = \pi r^2$ and then obtain $r = \sqrt{\frac{A}{\pi}}$ from it.

Alternatively, you can deal with the problem by solving an equation, as follows:

Suppose you want to find the radius of a circle which has an area of 250 m².
Using $A = \pi r^2$, $\quad 3.142 \times r^2 = 250$
$$r^2 = \frac{250}{3.142} = 79.567\ldots$$
$$r = \sqrt{79.567\ldots} = 8.9200\ldots$$

The radius of the circle is 8.92 m to 3 significant figures.

You can 'change the subject' for other mensuration formulae. For example, $C = 2\pi r$ for the circumference of a circle, can be changed to $r = \frac{C}{2\pi}$.

Area = length × breadth for a rectangle can be changed to

breadth = $\frac{\text{area}}{\text{length}}$

and $A = \frac{1}{2}bh$ for the area of a triangle can be changed to $h = \frac{2A}{b}$.

However, if you find manipulation of formulae difficult, you may prefer to use the 'solve an equation' method. This is the method used in the following examples.

Example 1

The area of a rectangular field is 2700 m².
If the length is 60 m, calculate the width.

Solution

For a rectangle, area = length × width
so 2700 = 60 × width

Dividing both sides by 60, $\frac{2700}{60}$ = width

Hence, the width of the field is 45 m.

Example 2

A rectangle is twice as long as it is wide. Its width is W cm.
a) Write down an expression for the length of the rectangle in terms of W.
b) The area of the rectangle is A cm².
    (i) Write down a formula for A, in terms of W.
    (ii) If $A = 128$, find the value of W.

Solution

a) For this rectangle, length = 2 × width
        so length = $2W$

b) (i) Area of rectangle = length × width
            so $A = 2W \times W$
            $A = 2W^2$
    (ii) If $A = 128$, $2W^2 = 128$
            $W^2 = 64$
            $W^2 = 8$

> IGCSE examination questions often test more than one part of the syllabus. This question tests knowledge of algebra as well as knowledge of mensuration. You must be prepared to tackle questions like this.

Example 3

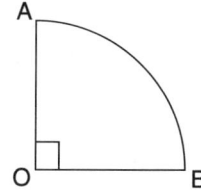

The diagram represents a quarter circle OAB. Its area is 16.5 cm²

Calculate the length of the radius OA.

Solution

The area of a circle is πr^2
so the area of a quarter circle is $\frac{1}{4}\pi r^2$
The area is 16.5 cm², so $\frac{1}{4}\pi r^2 = 16.5$
$$r^2 = \frac{4 \times 16.5}{3.142} = 21.006$$
$$r = \sqrt{21.006} = 4.583\ldots$$

The length of the radius is 4.58 cm correct to 3 significant figures.

Example 4

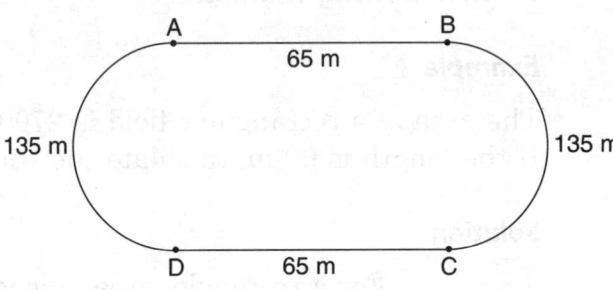

The diagram shows the plan of a running track. The two straights, AB and CD, are each 65 m long. The curved ends are both semicircles and each is 135 m long.

a) (i) Find the total distance, in metres, round 1 lap of the track.
    (ii) How many laps would have to be run in a 2 km race?
b) Find the radius, in metres, of the curved ends of the track.

Solution

a) (i) Perimeter of the track $= (65 + 135 + 65 + 135)$ m
                                $= 400$ m
    (ii) 2 km = 2000 m
         so number of laps in 2 km race $= \frac{2000}{400}$
         $= 5$
b) The semicircular ends each have a length of 135 m,
    so the circumference of the circle = 270 m.
    Hence, $2\pi r = 270$
    $$r = \frac{270}{2\pi} = 42.966\ldots$$

The radius of the curved ends is 43.0 m to 3 significant figures.

Example 5

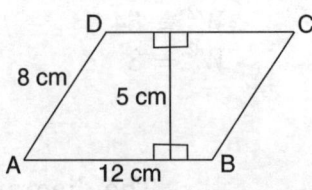

ABCD is a parallelogram in which AB = 12 cm and AD = 8 cm.

The perpendicular distance between AB and DC is 5 cm.

Calculate the perpendicular distance between AD and BC.

Solution

Using AB as base, we obtain
area of parallelogram ABCD = base × perpendicular height
$$= (12 \times 5) \text{ cm}^2 = 60 \text{ cm}^2$$

Using AD as base, we obtain
area of ABCD = 8 × perpendicular distance between AD and BC
and hence 60 = 8 × perpendicular distance between AD and BC
Thus, perpendicular distance between AD and BC = 7.5 cm.

Here are some questions for you to solve. Draw yourself a sketch and think before you calculate!

EXERCISE 7

1. A parallelogram ABCD has side AB = 8.2 cm and an area of 28.7 cm². Calculate the perpendicular distance between AB and DC.

2. Calculate, correct to 2 significant figures, the radius of a circle which has an area of 69.4 m².

3. A rectangle has a length of 5 m and breadth of 2 m.
    a) (i) Calculate the perimeter of the rectangle.
       (ii) Calculate the area of a square which has the same perimeter as the rectangle.
    b) (i) Calculate the area of the rectangle.
       (ii) Calculate the side of a square which has the same area as the rectangle.

4. a) Calculate the circumference of a circle of diameter 30 cm.

    b) The circumference of a trundle wheel is 100 cm.

    Work out the diameter of the trundle wheel correct to the nearest centimetre.

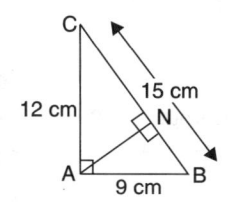

5. In the triangle ABC, angle A = 90°, AB = 9 cm, BC = 15 cm and CA = 12 cm.
    a) Calculate the area of the triangle ABC.
    b) Calculate the perpendicular distance, AN, from A to BC.

Check your answers at the end of this module.

If you are following the CORE syllabus, you have now completed your work for this unit and you should turn to the 'Summary' on page 41. If you are following the EXTENDED syllabus, there are a few more topics you need to study.

Perimeters and areas of similar figures

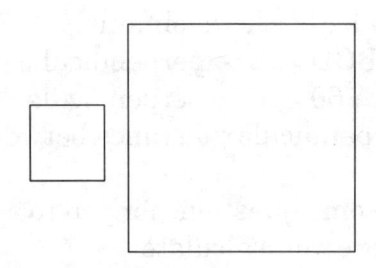

Look at these two squares. Is it true to say that one square is three times as big as the other?

The sides of the squares are 1 cm and 3 cm. The diagonals are 1.4 cm and 4.2 cm. The perimeters are 4 cm and 12 cm.

These facts seem to show that one square is 3 times as big as the other.

However, the areas are 1 cm² and 9 cm², so in fact the large square is *9* times as big as the small square!

This example shows that you must be careful when comparing sizes of shapes. The sides, diagonals and perimeters are all lengths. As far as lengths are concerned, the large square is 3 times as big as the small square. We call the ratio of corresponding sides the **linear scale factor**. For these two squares, the linear scale factor is 3. The areas are in the ratio 9 : 1 and we say that the **area scale factor** is 9. (Notice that 9 = 3 *squared* and that area is measured in *square* units.)

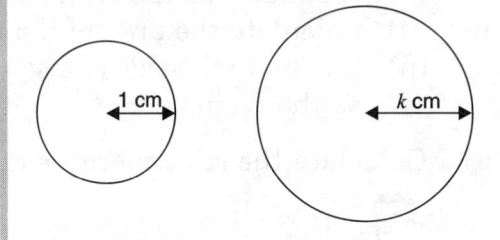

Look at these two circles.

The radii are 1 cm and k cm.
The diameters are 2 cm and $2k$ cm.
The circumferences are 4π cm and $4k\pi$ cm.

You can see that the **linear** scale factor is k.

However, the areas are π^2 and πk^2, so the **area** scale factor is k^2.

In fact, we can prove that:

> if the **linear** scale factor for a pair of similar shapes is k, then the **perimeter** scale factor is k and the **area** scale factor is k^2.

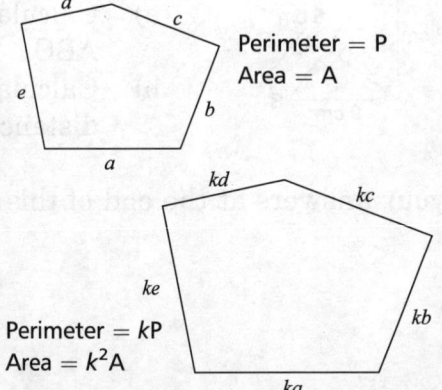

These results are fairly obvious for polygons. Suppose the linear scale factor is k. If the sides of the first polygon are $a, b, c, \ldots$, then the corresponding sides of the other polygon must be $ka, kb, kc, \ldots$

The perimeters are $a + b + c + \ldots$ and $ka + kb + kc + \ldots$ That is $(a + b + c + \ldots)$ and $k(a + b + c + \ldots)$. So the perimeter scale factor is k.

To obtain the result for areas, you regard the second polygon as an enlargement of the first polygon, with enlargement factor k. Every unit square (1 unit by 1 unit) inside the first polygon becomes a square k units by k units inside the second polygon. Thus, for every unit of area inside the first polygon, there are k units of area inside the second polygon. It follows that the area of the second polygon is k^2 times the area of the first polygon.

Example 1

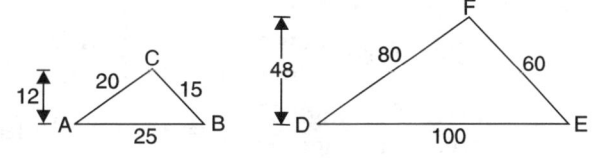

In the diagram, the dimensions of the triangles ABC and DEF are in millimetres.

a) Explain why triangles ABC and DEF are similar.
b) Find the perimeter of each triangle and the ratio

$$\frac{\text{perimeter of triangle DEF}}{\text{perimeter of triangle ABC}}.$$

c) Find the area of each triangle and the ratio

$$\frac{\text{area of triangle DEF}}{\text{area of triangle ABC}}.$$

Solution

a) Each side of the triangle DEF is 4 times the corresponding side of triangle ABC. Since the ratio of corresponding sides is constant, the triangles are similar.

b) Perimeter of triangle DEF = $(100 + 60 + 80)$ mm = 240 mm
Perimeter of triangle ABC = $(25 + 15 + 20)$ mm = 60 mm

$$\frac{\text{perimeter of triangle DEF}}{\text{perimeter of triangle ABC}} = \frac{240}{60} = 4$$

c) Area of triangle DEF = $\frac{1}{2}(100 \times 48)$ mm^2 = 2400 mm^2

Area of triangle ABC = $\frac{1}{2}(25 \times 12)$ mm^2 = 150 mm^2

$$\frac{\text{area of triangle DEF}}{\text{area of triangle ABC}} = \frac{2400}{150} = 16$$

In this example, the linear scale factor is 4 and the area scale factor is 16.
$16 = 4^2$, so we see that area scale factor = (linear scale factor)^2.

Example 2

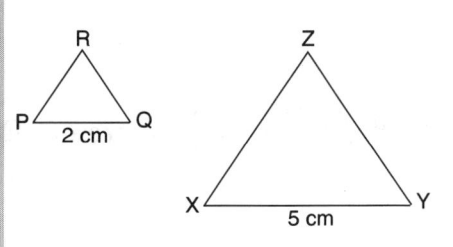

Triangles PQR and XYZ are equilateral. PQ = 2 cm and XY = 5 cm

a) Find the ratio

$$\frac{\text{perimeter of triangle XYZ}}{\text{perimeter of triangle PQR}}.$$

b) Find the ratio

$$\frac{\text{height of triangle XYZ}}{\text{height of triangle PQR}}.$$

c) Find the ratio

$$\frac{\text{area of triangle XYZ}}{\text{area of triangle PQR}}.$$

Solution

a) Perimeter of triangle XYZ = (5 + 5 + 5) cm = 15 cm
Perimeter of triangle PQR = (2 + 2 + 2) cm = 6 cm
$\frac{\text{perimeter of triangle XYZ}}{\text{perimeter of triangle PQR}} = \frac{15}{6} = 2.5$ [this ratio is equal to the ratio of corresponding sides XY and PQ]

b) The angles of triangle PQR are each 60° and the angles of triangle XYZ are each 60°.
Hence triangle XYZ is similar to triangle PQR (corresponding angles equal).
Therefore, $\frac{\text{height of triangle XYZ}}{\text{height of triangle PQR}} = $ linear scale factor = 2.5.

c) $\frac{\text{area of triangle XYZ}}{\text{area of triangle PQR}} = $ area scale factor = (linear scale factor)^2
$= (2.5)^2$
$= 6.25$

Example 3

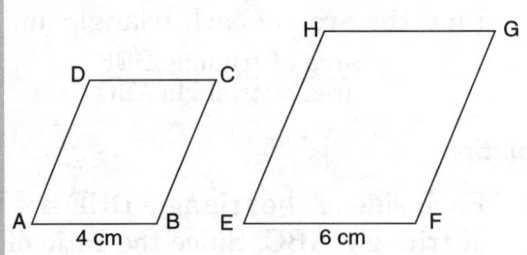

AB and EF are corresponding sides of two similar parallelograms ABCD and EFGH.
AB = 4 cm and EF = 6 cm.

a) Find the ratio $\frac{\text{perimeter of EFGH}}{\text{perimeter of ABCD}}$.

b) (i) Find the ratio $\frac{\text{area of EFGH}}{\text{area of ABCD}}$.

(ii) Given that the area of ABCD is 20 cm², find the area of EFGH.

Solution

The linear scale factor = ratio of corresponding sides = $\frac{6}{4} = \frac{3}{2}$

a) $\frac{\text{perimeter of EFGH}}{\text{perimeter of ABCD}} = $ linear scale factor $= \frac{3}{2}$

b) (i) $\frac{\text{area of EFGH}}{\text{area of ABCD}} = $ area scale factor = (linear scale factor)^2
$= (\frac{3}{2})^2$
$= \frac{9}{4}$

(ii) $\frac{\text{area of EFGH}}{20} = \frac{9}{4}$, so area of EFGH $= \frac{20 \times 9}{4}$ cm² = 45 cm²

Example 4

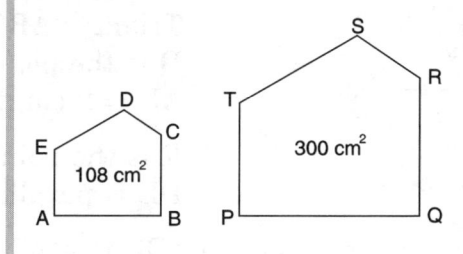

AB and PQ are corresponding sides of two similar pentagons ABCDE and PQRST.

PQ = 15 cm, area ABCDE = 108 cm^2 and area PQRST = 300 cm^2.

Calculate the length of AB.

Solution

Area scale factor = $\frac{\text{area of PQRST}}{\text{area of ABCDE}} = \frac{300}{108} = \frac{100}{36} = \frac{25}{9}$

(linear scale factor)^2 = area scale factor = $\frac{25}{9}$

linear scale factor = $\sqrt{\frac{25}{9}} = \frac{5}{3}$ and hence $\frac{PQ}{AB} = \frac{5}{3}$

Using PQ = 15 cm, we obtain $\frac{15}{AB} = \frac{5}{3}$ and so AB = $\frac{15 \times 3}{5}$ cm

That is AB = 9 cm.

Example 5

The scale of a map is 1 : 50 000.
On the map, the area representing a lake is 2 cm^2.
Calculate the actual area of the lake in square metres.

Solution

Linear scale factor = 50 000 = 5×10^4
so area scale factor = $(5 \times 10^4)^2 = 25 \times 10^8$

actual area of the lake = $25 \times 10^8 \times 2$ cm^2
= 50×10^8 cm^2

Since 1 m = 100 cm, 1 m^2 = 10 000 cm^2 = 10^4 cm^2

and so actual area of the lake = $\frac{(50 \times 10^8)}{10^4}$ m^2

= 50×10^4 m^2
= 500 000 m^2

Alternative method

Suppose the 2 cm^2 representing the lake is a rectangle 2 cm by 1 cm.
Using the scale 1:50 000 (which is the linear scale factor),
the actual lake is a rectangle 100 000 cm by 50 000 cm,
that is 1000 m by 500 m.
Actual area of the lake = (1000×500) m^2
= 500 000 m^2

Example 6

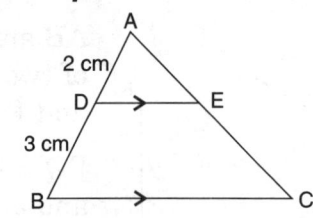

Triangle ABC has an area of 15 cm². D is the point on the side AB such that AD = 2 cm and DB = 3 cm.

E is the point on the side AC such that DE is parallel to BC.

a) Explain why triangle ADE is similar to triangle ABC.
b) Calculate the ratio $\frac{BC}{DE}$.
c) Calculate the area of the trapezium BCED.

Solution

a) In triangles ADE and ABC,
   angle ADE = angle ABC (corresponding angles, DE is parallel to BC)
   angle AED = angle ACB (corresponding angles, DE is parallel to BC)
   angle DAE = angle BAC (same angle)
   Hence, triangles ADE and ABC are equiangular and so they are similar.

b) $\frac{BC}{DE} = \frac{AB}{AD}$ (ratio of corresponding sides of similar triangles)
   $= \frac{2+3}{2}$
   $\frac{BC}{DE} = \frac{5}{2}$

c) For the similar triangles ADE and ABC, the linear scale factor $= \frac{5}{2}$.

   Hence, the area scale factor $= \frac{25}{4}$ and so $\frac{\text{area of triangle ABC}}{\text{area of triangle ADE}} = \frac{25}{4}$

   so area triangle ADE $= \frac{4 \times \text{area of triangle ABC}}{25} = \frac{4 \times 15}{25}$ cm²
   $= 2.4$ cm²

   hence, area trapezium BCED $= (15 - 2.4)$ cm² $= 12.6$ cm²

In this work, you must be careful to write the fractions (representing the ratios) the right way up. As you solve the problems in Exercise 8, you should be thinking "Should this fraction be less than 1 or greater than 1?"

EXERCISE 8

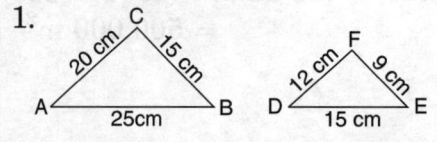

1. The length of the sides of triangle ABC are 15 cm, 20 cm and 25 cm. The lengths of the sides of triangle DEF are 9 cm, 12 cm and 15 cm.

a) Explain why triangle ABC is similar to triangle DEF.

b) Calculate the ratio $\frac{\text{perimeter of triangle ABC}}{\text{perimeter of triangle DEF}}$.

c) Calculate the ratio $\frac{\text{area of triangle ABC}}{\text{area of triangle DEF}}$.

2. In the rectangle JKLM, JK = 10 cm and KL = 15 cm.
   In the rectangle PQRS, PQ = 6 cm and QR = 9 cm.
   a) Explain why rectangle JKLM is similar to rectangle PQRS.
   b) Calculate the ratio $\frac{\text{perimeter of rectangle JKLM}}{\text{perimeter of rectangle PQRS}}$.
   c) Calculate the ratio $\frac{\text{area of rectangle JKLM}}{\text{area of rectangle PQRS}}$.

3. ABCD and EFGH are similar rhombuses.
   AB = 25 mm and EF = 45 mm.
   a) Calculate the ratio $\frac{EG}{AC}$.
   b) Calculate the ratio
      $\frac{\text{perimeter of rhombus EFGH}}{\text{perimeter of rhombus ABCD}}$.
   c) Calculate the ratio
      $\frac{\text{area of rhombus EFGH}}{\text{area of rhombus ABCD}}$.

4. The areas of two similar quadrilaterals are 72 cm² and 128 cm². The longest side of the first quadrilateral is 9 cm. Calculate the length of the longest side of the second quadrilateral.

5. The scale of a map is 1:10 000. On the map, the area representing a forest is 8 cm². Calculate the actual area of the forest in square metres.

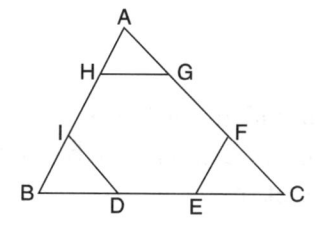

6. In the diagram, D and E are the points of trisection of BC. That means BD = DE = EC. F and G are the point of trisection of CA. H and I are the points of trisection of AB.

   Prove that the area of hexagon DEFGHI is two thirds of the area of triangle ABC.

Check your answers at the end of this module.

D Arcs and sectors of circles

As you know, a semicircle is half a circle and, in some of your previous work, you have taken the length of a semicircle to be half the circumference of the circle, and the area of a semicircle to be half the area of the circle. In this section, we extend the work to other fractions of a circle.

Length of a circular arc

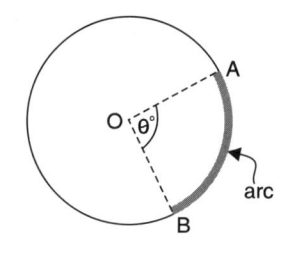

An arc is part of the circumference of a circle. The diagram shows an arc AB. The length of the arc depends on the angle ($\theta°$) which it subtends at the centre O of the circle.

The complete angle at O = 360°. If this angle is divided into 360 equal parts of 1°, each part is subtended by an arc which is $\frac{1}{360}$ of the circumference of the circle. (This is because the circle has rotational symmetry about its centre.)

It follows that the length of the arc AB is $\frac{\theta}{360}$ of the circumference.

Hence, we have the formula:

> length of an arc = $\frac{\theta}{360} \times 2\pi r$ where $\theta°$ = angle subtended by the arc at the centre of the circle

Area of a sector

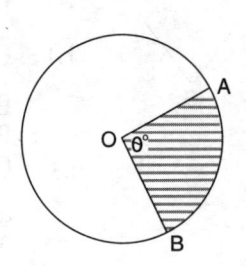

The region bounded by an arc AB of a circle (with centre O) and the two radii OA and OB is called a **sector** of the circle. The angle AOB is the 'angle of the sector'.
If $\theta° = 180°$, the sector is a half circle – that is a **semicircle**.
If $\theta° = 90°$, the sector is a quarter circle – that is a **quadrant**.

Using a similar argument as we used for the arc length, we find that the area of a sector is $\frac{\theta}{360}$ of the area of the circle.

Hence, we have the formula:

> area of a sector = $\frac{\theta}{360} \times \pi r^2$ where $\theta°$ is the angle of the sector

Example 1

A circle has a radius of 10 cm.
For a sector of this circle with 200°, calculate:
a) the perimeter   b) the area

Solution

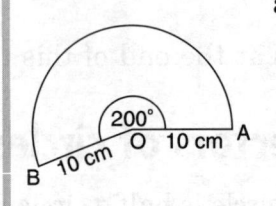

a) arc AB = $\frac{200}{360} \times 2\pi r = \frac{5}{9} \times 2 \times 3.142 \times 10$

arc AB = 34.91111 cm
perimeter = arc AB + radius OA + radius OB
= (34.91111 + 10 + 10) cm
= 54.91111 cm
= 54.9 cm to 3 significant figures

b) area of sector OAB = $\frac{200}{360} \times \pi r^2$

$= \frac{5}{9} \times 3.142 \times (10)^2 \text{ cm}^2$

= 174.55555 cm^2
= 175 cm^2 to 3 significant figures

Example 2

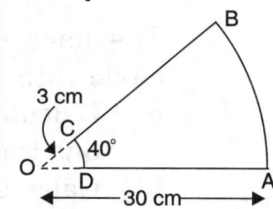

The diagram shows a blade ABCD of an electric fan. AB and CD are arcs of circles with centre O.

a) Calculate the perimeter of the blade.
b) Calculate the area of the blade.

Solution

a) arc AB $= \frac{40}{360} \times (2\pi r) = \frac{1}{9} \times 2 \times 3.142 \times 30$ cm $= 20.94666$ cm

arc CD $= \frac{40}{360} \times (2\pi r) = \frac{1}{9} \times 2 \times 3.142 \times 3$ cm $= 2.094666$ cm

BC = DA = (30 − 3) cm = 27 cm

perimeter of the blade = arc AB + arc CD + BC + DA
$= (20.94666 + 2.094666 + 27 + 27)$ cm
$= 77.041\ldots$ cm
$= 77.0$ cm to 3 significant figures

b) area of sector OAB $= \frac{40}{360} \times \pi r^2 = \frac{1}{9} \times 3.142 \times (30)^2$ cm^2
$= 314.2$ cm^2

area of sector OCD $= \frac{40}{360} \times \pi r^2 = \frac{1}{9} \times 3.142 \times (3)^2$ cm^2
$= 3.142$ cm^2

area of the blade $= (314.2 - 3.142)$ cm$^2 = 311.058$ cm^2
$= 311$ cm^2 to 3 significant figures

Example 3

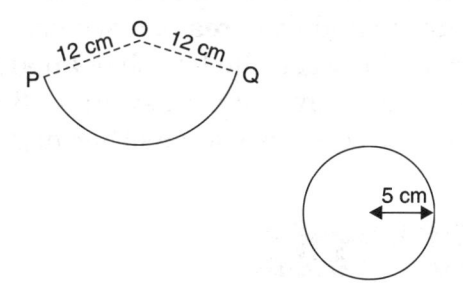

The diagram shows a sector OPQ of a circle with radius 12 cm and centre O.

The length of the arc PQ is the same as the circumference of a circle with radius 5 cm.

a) Calculate the angle POQ.
b) Calculate the area of the sector OPQ.

Solution

a) Length of arc PQ = circumference of circle with radius 5 cm
$= 2 \times \pi \times 5$ cm
$= 10\pi$ cm

If angle POQ $= \theta$, length of arc PQ $= \frac{\theta}{360} \times 2\pi \times 12$ cm $= \frac{\theta \pi}{15}$ cm

Hence, $\frac{\theta \pi}{15} = 10\pi$ and so $\theta = \frac{15 \times 10\pi}{\pi} = 150$

angle POQ $= 150°$

b) area of sector OPQ $= \frac{150}{360} \times \pi r^2 = \frac{5}{12} \times 3.142 \times (12)^2$ cm^2

area of sector OPQ = 188.52 cm^2 = 189 cm^2 to 3 significant figures

The region between an arc and a chord is called a **segment** of the circle.

Example 4

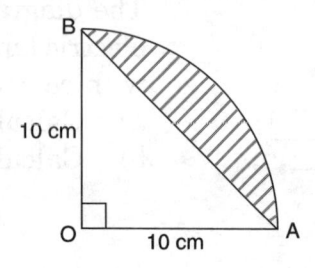

The diagram shows a quadrant OAB of a circle with centre O and radius 10 cm.
a) Calculate the perimeter of the quadrant.
b) Calculate the area of the quadrant.
c) Calculate the area of the region between the arc AB and the chord AB.

Solution

a) arc AB = $\frac{90}{360} \times 2\pi r = \frac{1}{4} \times 2 \times 3.142 \times 10$ cm = 15.70 cm

   perimeter of quadrant = (15.70 + 10 + 10) cm = 35.70 cm
   perimeter of quadrant = 35.7 cm to 3 significant figures

b) area of quadrant = $\frac{90}{360} \times \pi r^2 = \frac{1}{4} \times 3.142 \times (10)^2$ cm^2
   = 78.55 cm^2
   area of quadrant = 78.6 cm^2 to 3 significant figures

c) area of triangle OAB = $\frac{1}{2}$(base × height) = $\frac{1}{2}(10 \times 10)$ cm^2
   = 50 cm^2

   area of shaded segment = area of quadrant − area of triangle OAB
   = (78.55 − 50) cm^2 = 28.55 cm^2
   area of shaded segment = 28.6 cm^2 to 3 significant figures

You should now be able to use the formulae for the length of a circular arc and the area of a sector. You should also be aware of the difference between the length of an arc and the perimeter of a sector. Remember to give your answers to 3 significant figures unless you are instructed to use some other degree of accuracy.

EXERCISE 9

1. A circle has a radius of 12 cm.
   For a sector of this circle with angle 50°, calculate:
   a) the perimeter
   b) the area

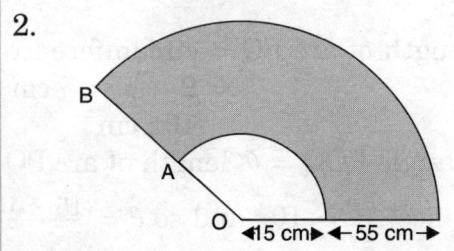

2. The diagram shows the region of a windscreen which is wiped by a rubber blade AB which is 55 cm long. The end A moves along an arc of a circle of radius 15 cm and centre O. The arc subtends an angle of 120° at O.

   Calculate the area of the region wiped by the blade.

3. A circle has a radius of 10 cm.

Calculate, correct to the nearest degree, the angle subtended at the centre by an arc which is 14 cm long.

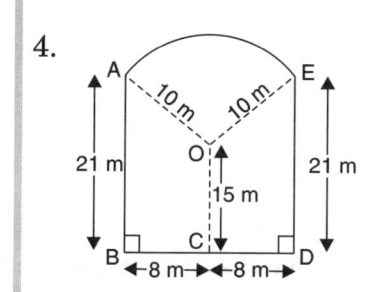

4. The diagram represents the floor of a room. OABC and OCDE are trapeziums. OAE is a sector of a circle with centre O, and angle AOE = 106°.
   a) Calculate the perimeter of the floor.
   b) Calculate the area of the floor.

Check your answers at the end of this module.

Summary

In this unit you learnt that to calculate the perimeter of a polygon you add up the lengths of the sides. You also learnt a few formulae for calculating the areas of polygons:

- area of a rectangle = length × breadth
- area of a triangle = $\frac{1}{2}$ base × perpendicular height
- area of a parallelogram = base × perpendicular height
- area of a trapezium = average of the parallel sides × perpendicular distance between.

In the section on circles you learnt:

- circumference of a circle = $\pi d = 2\pi r$
- area of a circle = πr^2.

Those of you studying the EXTENDED syllabus also learnt the following regarding the perimeters and areas of similar figures:

- if the linear scale factor for a pair of similar shapes is k, then the perimeter scale factor is k and the area scale factor is k^2.

You also saw that:

- the length of an arc of a circle = $\frac{\theta}{360} \times 2\pi r$ where θ = angle subtended by the arc at the centre of the circle
- the area of a sector of a circle = $\frac{\theta}{360} \times \pi r^2$ where θ is the angle of the sector.

In Unit 2 you will be learning about how to calculate the surface areas and volumes of 3-dimensional objects. Before you continue with Unit 2, do the 'Check your progress' on the next page.

Check your progress

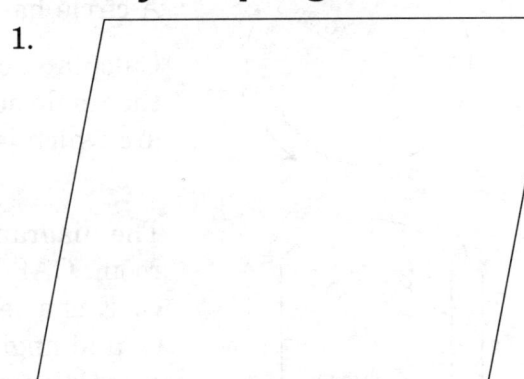

1. The diagram shows a parallelogram which is drawn full size.

   After making the necessary measurements, calculate
   a) the perimeter
   b) the area of the parallelogram

   (Each of the measurements you require is a whole number of centimetres.)

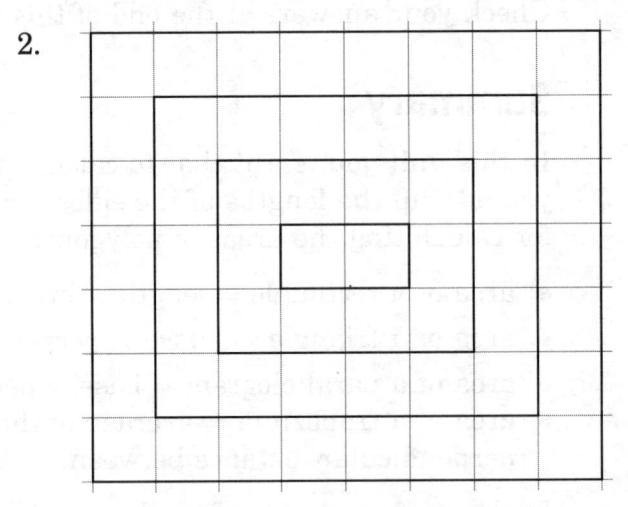

2. Four rectangles are outlined in the diagram.

   The area of the innermost one is 2 cm² and its perimeter is 6 cm.
   a) Working outwards, write down the area and perimeter of the other three rectangles.

Rectangle	Area (cm^2)	Perimeter (cm)
1st	$1 \times 2 = 2$	$2(1 + 2) = 6$
2nd	$3 \times 4 = \ldots$	$2(3 + 4) = \ldots$
3rd	$\ldots \times \ldots = \ldots$	$2(\ldots + \ldots) = \ldots$
4th	$\ldots \times \ldots = \ldots$	$2(\ldots + \ldots) = \ldots$

   b) Write down the area and perimeter of the next two rectangles in the sequence (not shown in the diagram).

   c) Write down the area and perimeter of the fifteenth rectangle in the sequence.

3. Car tyres need replacing after 7.5×10^6 revolutions.
   If the wheel radius of a car is 0.3 m, calculate the distance, in metres, the car can travel before its tyres need replacing.

4. A piece of rope is 12 m long. It is laid on the ground in a circle, as shown in the diagram.
   a) Calculate the diameter of the circle.
   b) The cross-section of the rope is a circle of radius 1.2 cm. Calculate the area of the cross-section.

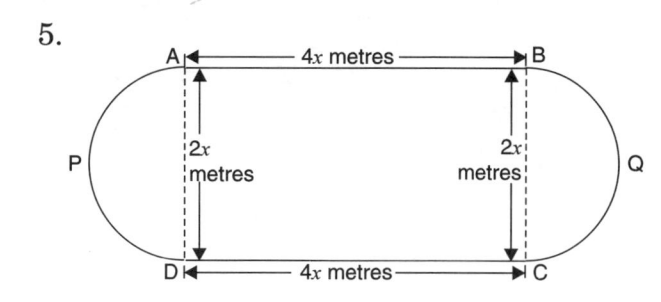

5. The diagram shows a running track, consisting of two parallel lines, AB and DC, each of length $4x$ metres, and two semicircles, BQC and APD, each of diameter $2x$ metres.

   a) (i) Write down an expression, in terms of x and π, for the circumference of a circle of diameter $2x$ metres.
      (ii) Write down an expression, in terms of x and π, for the perimeter ABQCDPA of the running track.
      (iii) Factorise completely your answer to part a) (ii).

   b) The perimeter of the running track is 400 m.
      (i) Use your answer to part a) (iii) to show that $(\pi + 4)x = 200$.
      (ii) Taking π to be 3.142, solve the equation $(\pi + 4)x = 200$.
      (iii) Write down the lengths of AB, BQC, CD, DPA.

6. A circle has centre O and a radius of 12 cm.
   A sector OAB of this circle has an area of 60π cm^2.
   a) Calculate the size of angle AOB.
   b) Calculate, in terms of π, the perimeter of sector OAB.

Check your answers at the end of this module.

Unit 2
Surface Area and Volume

In Unit 1 you studied shapes drawn on a flat surface (2-dimensional shapes). In this unit you'll be working with **solid** shapes (3-dimensional shapes).

This unit is divided into four sections:

Section	Title	Time
A	Surface area of common solids	2 hours
B	Volume of common solids	3 hours
C	Extending the use of the formulae	2 hours
D	Surface area and volume of pyramids, cones and spheres	3 hours

By the end of this unit, you should be able to:

- calculate the surface area of cuboids and cylinders
- calculate the volume of cuboids, prisms and cylinders
- use the relationships between units of length, area and volume
- solve problems involving surface area and volume of solids
- use the formulae for the surface area and volume of pyramids, cones and spheres
- use the relationship between the surface areas of similar solids and the volumes of similar solids.

A Surface area of common solids

The boundary of a 2-dimensional shape consists of lines which may be straight or curved. The total length of the boundary is called the perimeter. The boundary of a solid consists of surfaces (called **faces** which may be plane (flat) or curved. The total area of these faces is called the **surface area** of the solid.

You already know that area is measured in square units. The most common units for surface area are square millimetres (mm^2), square centimetres (cm^2) and square metres (m^2).

If all the faces of a solid are plane, then the solid is a **polyhedron** and each face is a **polygon**. You know how to find the area of a polygon, so you will be able to find the surface area of a polyhedron.

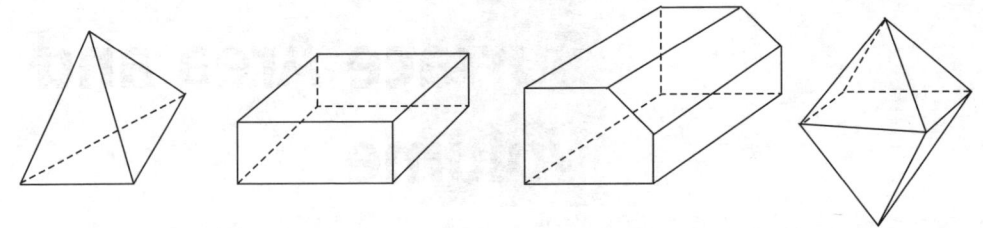

It is difficult to draw a 2-dimensional picture of a polyhedron. Most (if not all) of the faces will be distorted. A rectangle will not usually look like a rectangle, an equilateral triangle will not usually look like an equilateral triangle, etc. You will have to be careful when calculating the surface area of the solid.

If you can draw a net of the polyhedron, then each face is shown with the right shape and size. The area of the net is the same as the surface area of the solid. You will probably find it easy to calculate the area of the net. In most cases the net consists of rectangles and triangles.

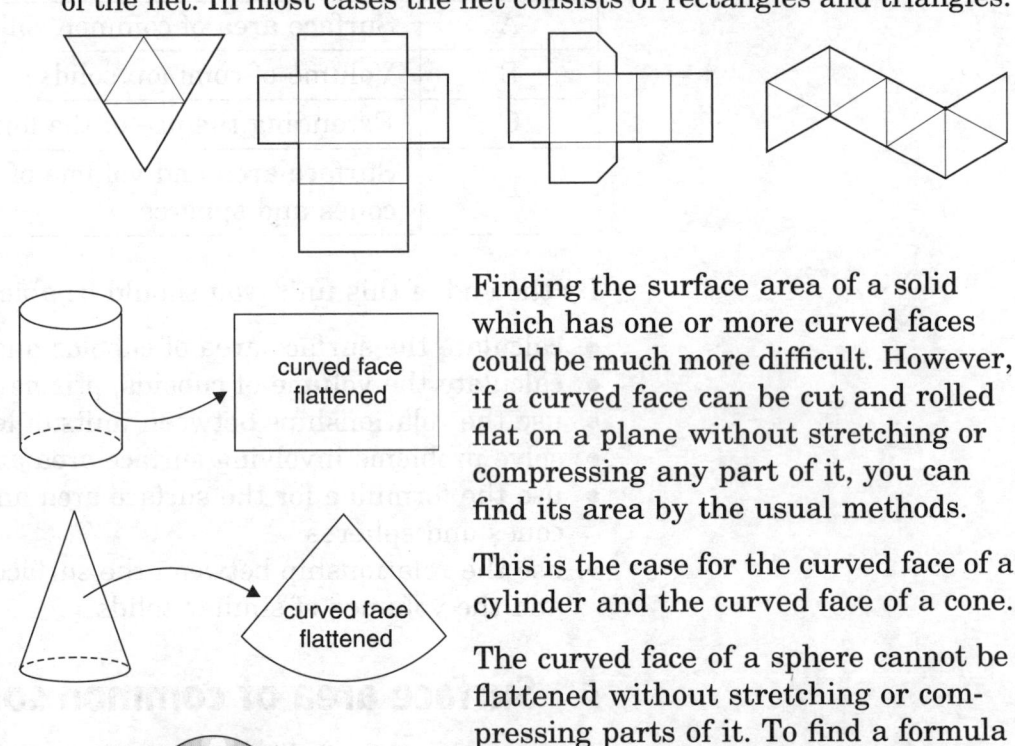

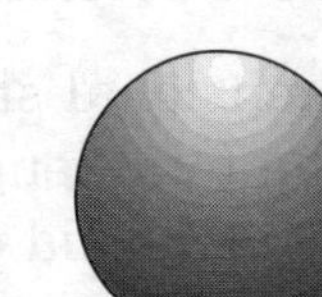

Finding the surface area of a solid which has one or more curved faces could be much more difficult. However, if a curved face can be cut and rolled flat on a plane without stretching or compressing any part of it, you can find its area by the usual methods.

This is the case for the curved face of a cylinder and the curved face of a cone.

The curved face of a sphere cannot be flattened without stretching or compressing parts of it. To find a formula to calculate the surface area of a sphere it is necessary to use more advanced mathematical methods.

Surface area of a cuboid

For the IGCSE CORE syllabus, you have to be able to find the surface area of cuboids and cylinders.

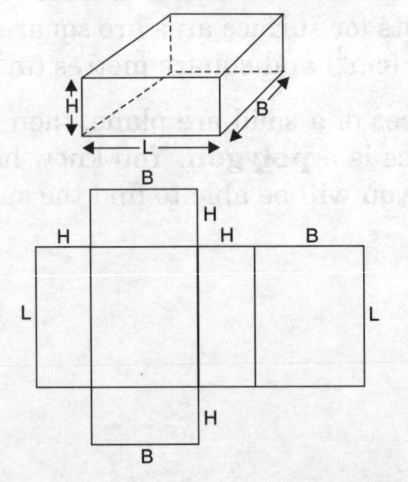

A cuboid has six faces and each of them is a rectangle. If the length, breadth and height of the cuboid are L units, B units and H units, respectively, then the surface area is A square units, where

$$A = 2LB + 2BH + 2HL.$$

You need not try to remember this formula. For any particular cuboid, you should calculate the area of each face and add the areas up, or you should find the area of the net.

Example 1

1. Find the surface area of a cuboid with:
   a) length = 12 cm, breadth = 9 cm and height = 5 cm
   b) length = 8 cm, breadth = 7 cm and height = 5 mm

Solution

a) There are two faces each with area (12×9) cm^2
   two faces each with area (9×5) cm^2
   and two faces each with area (5×12) cm^2

   Total surface area $= (108 \times 2)$ cm$^2 + (45 \times 2)$ cm$^2 + (60 \times 2)$ cm^2
   $= (216 + 90 + 120)$ cm^2
   $= 426$ cm^2

b) There is a mixture of units here (cm and mm).
   We must use the fact that 1 cm = 10 mm to change the dimensions to one of these units. We will change 5 mm to 0.5 cm and obtain an answer in cm^2.
   (We could have changed 8 cm to 80 mm and 7 cm to 70 mm and obtained an answer in mm^2.)

   Total surface area $= (8 \times 7)$ cm$^2 \times 2 + (7 \times 0.5)$ cm$^2 \times 2 + (0.5 \times 8)$ cm$^2 \times 2$
   $= (112 + 7 + 8)$ cm^2
   $= 127$ cm^2

Example 2

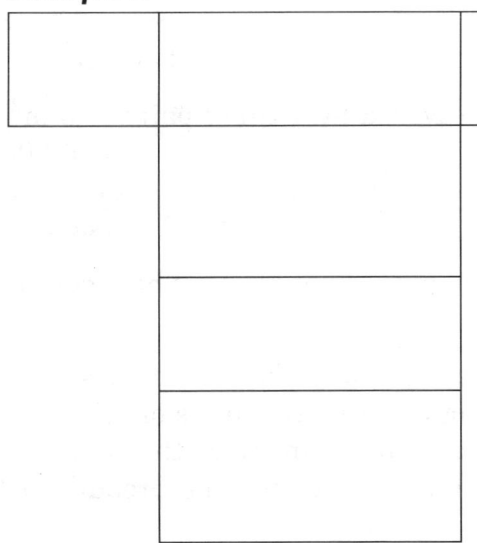

This is the net of a cuboid, drawn to scale.
a) Sketch the cuboid and, on the sketch, show clearly the length, breadth and height of the cuboid.
b) Find the surface area of the cuboid.

Solution

a) This is a sketch of the cuboid. The length, breadth and height are 4 cm, $1\frac{1}{2}$ cm and 2 cm. (These could be in any order. They were obtained by measuring the net.)

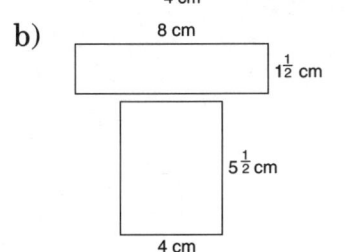

b) Surface area of the cuboid
   = area of the net
   $= (8 \times 1\frac{1}{2})$ cm$^2 + (5\frac{1}{2} \times 4)$ cm^2
   $= (12 + 22)$ cm^2
   $= 34$ cm^2

> **Rectangular box** is another way of saying **cuboid**.

Example 3

A room has the shape of a rectangular box. The floor is 5 m by 4 m and the height of the room is 2.4 m. The walls and ceiling of the room are to be sound-proofed. Calculate the total area of the walls and ceiling.

Solution

Notice that we do not have to calculate the *total* surface area of the rectangular box. The floor is not to be sound-proofed.

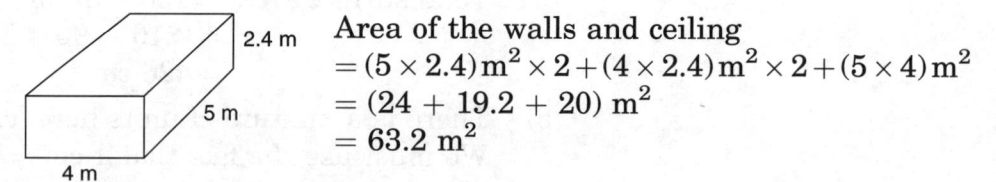

Area of the walls and ceiling
$= (5 \times 2.4)\,\text{m}^2 \times 2 + (4 \times 2.4)\,\text{m}^2 \times 2 + (5 \times 4)\,\text{m}^2$
$= (24 + 19.2 + 20)\,\text{m}^2$
$= 63.2\,\text{m}^2$

Example 4

A child's building bricks are cuboids. Each brick is 8 cm by 6 cm by 5 cm. The bricks are to be hand-painted with lead-free paint. 1 tin of paint will cover an area of 5 m^2.
How many bricks can be painted with 1 tin of paint?

Solution

Surface area of 1 brick $= (8 \times 6)\,\text{cm}^2 \times 2 + (6 \times 5)\,\text{cm}^2 \times 2 + (5 \times 8)\,\text{cm}^2 \times 2$
$= (96 + 60 + 80)\,\text{cm}^2$
$= 236\,\text{cm}^2$

Area covered by 1 tin of paint $= 5\,\text{m}^2$
$= 50\,000\,\text{cm}^2$

> remember 1 square metre = 100 cm × 100 cm = 10 000 cm^2

$\dfrac{\text{area covered by 1 tin of paint}}{\text{surface area of 1 brick}} = \dfrac{50\,000\,\text{cm}^2}{236\,\text{cm}^2} = 211.86$

Hence, number of bricks which can be painted with 1 tin of paint $= 211$.

I hope you found these examples easy to follow – basically they are all about finding the areas of rectangles. Now try some questions for yourself – but remember that to find the area of a rectangle, you must have the length and breadth *in the same units*.

EXERCISE 10

1. Find the surface area of a cuboid with:
    a) length = 56 mm, breadth = 45 mm and height = 36 mm
    b) length = 1.2 m, breadth = 75 cm and height = 60 cm

Module 5 Unit 2

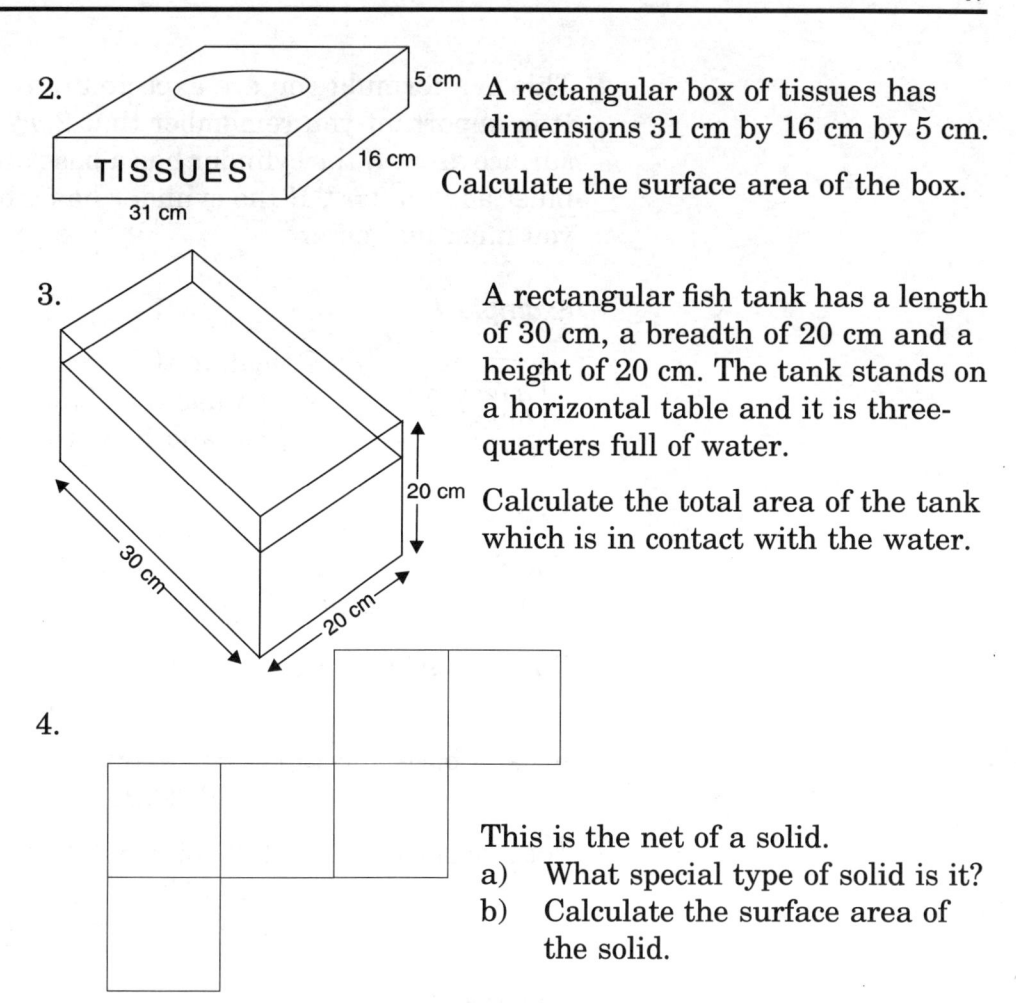

2. A rectangular box of tissues has dimensions 31 cm by 16 cm by 5 cm.

Calculate the surface area of the box.

3. A rectangular fish tank has a length of 30 cm, a breadth of 20 cm and a height of 20 cm. The tank stands on a horizontal table and it is three-quarters full of water.

Calculate the total area of the tank which is in contact with the water.

4. This is the net of a solid.
a) What special type of solid is it?
b) Calculate the surface area of the solid.

Check your answers at the end of this module.

Surface area of a cylinder

The surface of a cylinder consists of a flat circular base, a flat circular top and one curved face.

The areas of the base and top are found by using the formula $A = \pi r^2$.

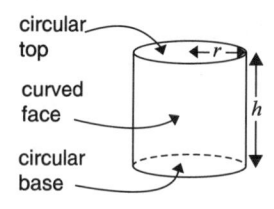

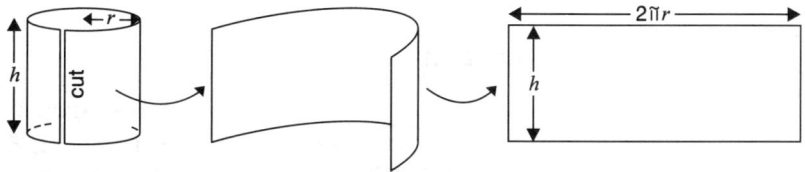

The curved face can be cut parallel to the axis of the cylinder and then flattened to make a rectangle.

The breadth of the rectangle is the same as the height (h) of the cylinder. The length of the rectangle is the same as the circumference of the base (or the top) of the cylinder, that is $2\pi r$.

$$\text{The area of the rectangle} = \text{length} \times \text{breadth} = 2\pi r \times h$$
$$= 2\pi rh.$$

It follows that:

> the *curved* surface area of a cylinder $= 2\pi rh$

This is a formula you are expected to remember.

It is important you remember that $2\pi rh$ gives only the *curved* surface area. If the cylinder has a base and a top (like a soup tin) you must add on $2\pi r^2$. If the cylinder has a base but no top (like a mug) you must add on πr^2.

Example 1

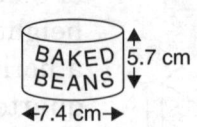

Calculate the surface area of a small tin of beans which has the shape of a cylinder with diameter 7.4 cm and height 5.7 cm.

Solution

The first step is to find the radius of the cylinder.

Radius $= \frac{\text{diameter}}{2} = \frac{7.4 \text{ cm}}{2} = 3.7$ cm

Curved surface area $= 2\pi rh$
$= 2 \times 3.142 \times 3.7 \times 5.7 \text{ cm}^2 = 132.52956 \text{ cm}^2$

Area of base and top $= \pi r^2 + \pi r^2 = 3.142 \times (3.7)^2 \times 2 \text{ cm}^2$
$= 86.02796 \text{ cm}^2$

Total surface area $= (132.52956 + 86.02796) \text{ cm}^2$
$= 218.55752 \text{ cm}^2$
$= 219 \text{ cm}^2$ to 3 significant figures

Example 2

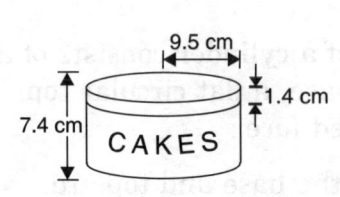

A food container, made of sheet metal, has the shape of a cylinder with radius 9.5 cm and height 7.4 cm.

The container has a lid, 1.4 cm deep, which fits tightly over the top.

Calculate the total area of sheet metal in the container (including the lid).

Solution

Surface area of lid $= \pi r^2 + 2\pi rh$
$= (3.142 \times (9.5)^2 + 2 \times 3.142 \times 9.5 \times 1.4) \text{ cm}^2$
$= (283.5655 + 83.5772) \text{ cm}^2$
$= 367.1427 \text{ cm}^2$

Surface area of lower part $= \pi r^2 + 2\pi rh$
$= (3.142 \times (9.5)^2 + 2 \times 3.142 \times 9.5 \times 7.4) \text{ cm}^2$
$= (283.5655 + 441.7652) \text{ cm}^2$
$= 725.3307 \text{ cm}^2$

Total area of sheet metal $= (367.1427 + 725.3307) \text{ cm}^2$
$= 1092.4734 \text{ cm}^2$
$= 1090 \text{ cm}^2$ to 3 significant figures

Example 3

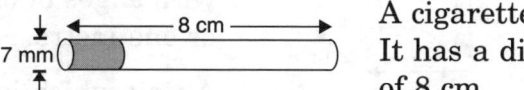

A cigarette is cylindrical in shape. It has a diameter of 7 mm and a length of 8 cm.

Calculate the area of the paper which forms the outside of the cigarette.

Solution

The paper has the form of the curved face of a cylinder – it does not have a base or a top.

We must find the radius, and the radius and height must be in the same units. Radius $= \frac{\text{diameter}}{2} = \frac{7 \text{ mm}}{2} = 3.5$ mm $= 0.35$ cm

Area of the paper $= 2\pi rh$
$= 2 \times 3.142 \times 0.35 \times 8$ cm^2
$= 17.5952$ cm^2
$= 17.6$ cm^2 to 3 significant figures

Here is a short exercise for you to do. Each question requires the use of the formula for the curved surface area of a cylinder. You must consider whether it is necessary to add on the area of the ends.

EXERCISE 11

1. Calculate the surface area of a cylindrical tank, closed at both ends, which has a radius of 25 cm and a height of 80 cm.

2. Calculate the surface area of a coin which has the shape of a cylinder with diameter 2.2 cm and thickness 3 mm.

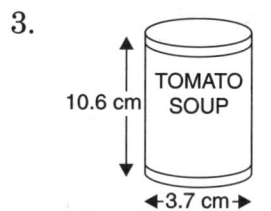

3. A cylindrical can of soup has a radius of 3.7 cm.

   A label, 10.6 cm wide, is wrapped round the can and has an overlap of 0.7 cm.

   Calculate the area of the label.

Check your answers at the end of this module.

B Volume of common solids

Measurement of volume

The **volume** of a solid (3-dimensional) shape is a measure of the amount of space it occupies. It corresponds to the area of a plane (2-dimensional) shape which you studied in Unit 1.

To measure the volume of a solid, we have to compare it with some unit of volume. This could be any shape which will fill space without leaving any gaps. In practice, we use the simplest possible shape – that is, a cube with edges of unit length.

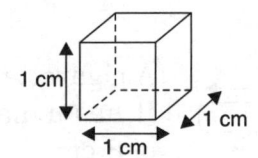

Depending on the situation, we choose cubes with edges of one millimetre, one centimetre or one metre.

A unit which is a cube with edges of one centimetre is called a **cubic centimetre**.

a cubic centimetre

Since a cubic centimetre has dimensions 1 cm by 1 cm by 1 cm, this unit is often abbreviated to 1 cm^3. read this as 'one centimetre cubed'

The units and abbreviations you need to know are shown below.

Unit	Abbreviation
cubic millimetre	mm^3
cubic centimetre	cm^3
cubic metre	m^3

Units of volume

The units of volume are related to one another and to the units of length, as shown in this table.

Length	Volume
1 cm = 10 mm	1 cm^3 = 10 mm × 10 mm × 10 mm = 1000 mm^3
1 m = 100 cm	1 m^3 = 100 cm × 100 cm × 100 cm = 1 000 000 cm^3

Volume of a cuboid

Suppose you have to find the volume of a cuboid which has a length of 4 cm, a breadth of 3 cm and a height of 2 cm.

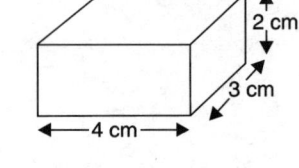

Start with a rectangular base, 4 cm by 3 cm. The area of this base is 12 cm^2 and so it is possible to cover it with 12 cubes, each with edges of 1 cm.

This is the bottom layer of the solid and it has a height of 1 cm.

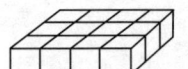

On top of each unit cube in the bottom layer, place another unit cube to form a second layer.

The solid now has a height of 2 cm and it contains (12 × 2) unit cubes, that is, 24 unit cubes.

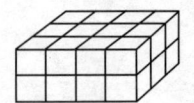

The cuboid with dimensions 4 cm by 3 cm by 2 cm has a volume of 24 cm^3.

If we now consider a cuboid with length L units, breadth B units and height H units (where L, B and H are whole numbers), the base is a rectangle which can be covered with (L × B) unit cubes. To build up the solid of height H units, we need H layers of (L × B) unit cubes, that is L × B × H unit cubes altogether.

When L, B and/or H are not whole numbers, we have to work with fractions (or decimals) of a unit cube, or use smaller cubes. It can be shown that for all values of the length, breadth and height:

$$\text{volume of a cuboid} = \text{length} \times \text{breadth} \times \text{height}$$

Example 1

Calculate the volume of a cuboid with:
a) length = 9.5 cm, breadth = 6.5 cm and height = 4 cm
b) length = $3\frac{3}{4}$ cm, breadth = $2\frac{2}{5}$ cm and height = $3\frac{1}{3}$ cm

Solution

a) Volume of cuboid = length × breadth × height
$= 9.5 \times 6.5 \times 4 \text{ cm}^3$
$= 247 \text{ cm}^3$

b) Volume of cuboid = length × breadth × height
$= 3\frac{3}{4} \times 2\frac{2}{5} \times 3\frac{1}{3} \text{ cm}^3$
$= \frac{15}{4} \times \frac{12}{5} \times \frac{10}{3} \text{ cm}^3$
$= 30 \text{ cm}^3$

Example 2

A rectangular paddling pool is 6 m by 4 m. In all parts of the pool the water is 40 cm deep. Calculate the volume of water in the pool.

Solution

Depth of water = 40 cm = 0.4 m

remember, the three dimensions must all be in the same units

Volume of water = length of pool × breadth of pool × depth of water
$= 6 \times 4 \times 0.4 \text{ m}^3$
$= 9.6 \text{ m}^3$

Example 3

A rectangular block of metal has dimensions 25 cm by 16 cm by 10 cm. It is melted down and recast into cubes with edge length 3 cm. How many cubes can be made?

Solution

Volume of metal = length × breadth × height
$= 25 \times 16 \times 10 \text{ cm}^3$
$= 4000 \text{ cm}^3$

Volume of one cube = $3 \times 3 \times 3 \text{ cm}^3$
$= 27 \text{ cm}^3$

$\frac{\text{volume of metal}}{\text{volume of cube}} = \frac{4000 \text{ cm}^3}{27 \text{ cm}^3} = 148.148$

Hence, number of cubes that can be made = 148

Example 4

Cornflake boxes have the shape of a cuboid with dimensions 25 cm by 20 cm by 6 cm.

How many of these cornflake boxes will fit into a carton which has the shape of a cuboid with dimensions 75 cm by 40 cm by 30 cm?

Solution

Dimensions   Dimensions
of carton   of a box
↓   ↓
75 cm   ÷   25 cm = 3
40 cm   ÷   20 cm = 2
30 cm   ÷   6 cm = 5

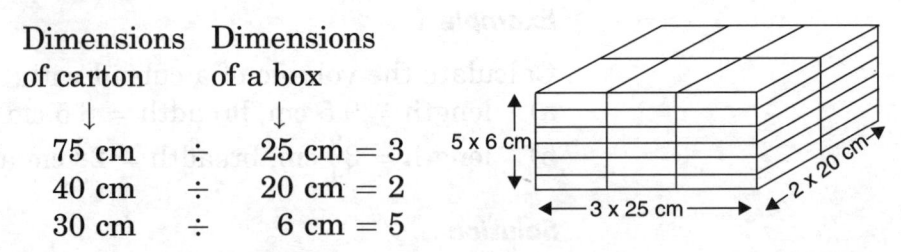

Number of boxes which can be fitted into the carton $= 3 \times 2 \times 5 = 30$.

The boxes will fit in the carton as shown in the diagram. The 30 boxes will fill the carton completely.

Warning: Do not use the method of Example 3 to solve questions like Example 4. With the dimensions given for the box and the carton, the method of Example 3 would give

$$\text{number of boxes} = \frac{\text{volume of carton}}{\text{volume of box}} = \frac{90\,000 \text{ cm}^3}{3000 \text{ cm}^3} = 30$$

Although this is the correct answer, the method used is incorrect. To understand this, look at Example 5.

Example 5

Muesli boxes have the shape of a cuboid with dimensions 20 cm by 15 cm by 4 cm.

How many of these boxes will fit into a carton which has the shape of a cuboid with dimensions 75 cm by 40 cm by 30 cm?

Solution

If you consider one end of the carton which is 40 cm by 30 cm, you can cover it with 4 boxes using their 20 cm by 15 cm faces. These boxes occupy 4 cm of the 75 cm length of the carton.
$75 \div 4 = 18$ with a remainder of 3, so you can fit in 18 layers of 4 boxes — that is 72 boxes altogether, and there will be a gap of 3 cm at one end of the carton.

The diagram shows how the boxes fit in. If you try any other way, you will find that you can fit in less than 72 boxes.

Using the method of Example 3, you would get

$$\text{number of boxes} = \frac{\text{volume of carton}}{\text{volume of box}} = \frac{90\,000 \text{ cm}^3}{1200 \text{ cm}^3} = 75$$

As we have seen, 75 is an incorrect answer!

EXERCISE 12

1. Each of the solids below is made up of unit cubes with edges of 1 cm. Find the volume of each solid.

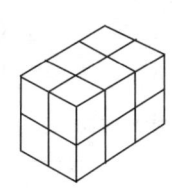

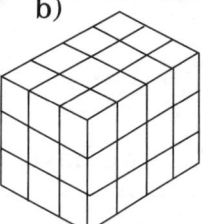

   a)   b)   c)

2. Calculate the volume of a cuboid with:
   a) length = 8.25 cm, breadth = 7.5 cm, height = 16 cm
   b) length = $4\frac{1}{2}$ cm, breadth = $3\frac{1}{3}$ cm, height = $1\frac{2}{5}$ cm

3. Calculate, in cubic centimetres, the volume of a rectangular sheet of glass, 80 cm by 75 cm, which is 3 mm thick.

4. A rectangular block of metal has dimensions 350 mm by 90 mm by 80 mm. It is machined to reduce it in size to 300 mm by 85 mm by 70 mm.
   a) Calculate the volume of the original block of metal.
   b) Calculate the volume of metal removed.

Check your answers at the end of this module.

Volume of a prism

For a cuboid, length × breadth gives the area of the base, and so the formula for its volume could be written as
   volume = area of base × height.

As we shall see, this formula applies to some other solids.

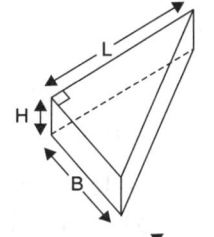

Consider the upright prism shown in the diagram.

Its cross-section is a right-angled triangle.

Two of these prisms can be put together to form a cuboid.

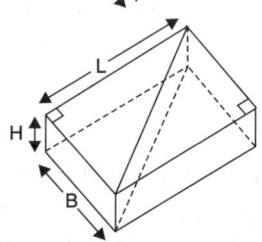

The volume of the cuboid = L × B × H, so the volume of each prism = $\frac{1}{2}$ × L × B × H.

$\frac{1}{2}$ × L × B = area of cross-section of prism, so volume of prism = area of cross-section × height.

If the cross-section of a triangular prism is not right-angled, the prism can be divided into two prisms which do have right-angled cross-sections, as shown in the diagram.

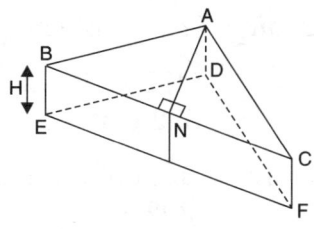

Volume of the triangular prism ABCDEF
= (area of triangle ABN) × H + (area of triangle ACN) × H
= (area triangle ABN + area triangle ACN) × H
= (total area of cross-section) × H

Now consider an upright prism which has a polygon as its cross-section. The polygon can be divided up into triangles and so the prism can be divided up into triangular prisms.

Suppose the areas of the triangles are $A_1, A_2, A_3, \ldots$

Then the volume of the prism $= (A_1 \times H) + (A_2 \times H) +$
$(A_3 \times H) + \ldots$
$= (A_1 + A_2 + A_3 + \ldots) \times H$
$= (\text{area of cross-section}) \times H$

Thus we have a formula which applies to all prisms:

> volume of a prism = area of cross-section × height

Example 1

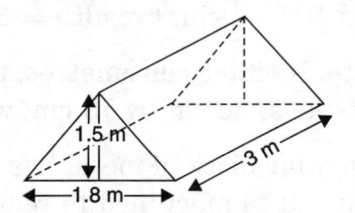

The diagram shows a ridge tent which has a length of 3 m.

Each end of the tent is an isosceles triangle with base 1.8 m and height 1.5 m.

Calculate the volume of the tent.

Solution

Area of cross-section $= \frac{1}{2} \times \text{base} \times \text{height}$
$= \frac{1}{2} \times 1.8 \times 1.5 \text{ m}^2$
$= 1.35 \text{ m}^2$

Volume of the tent $= \text{area of cross-section} \times \text{length}$
$= 1.35 \times 3 \text{ m}^3$
$= 4.05 \text{ m}^3$

Example 2

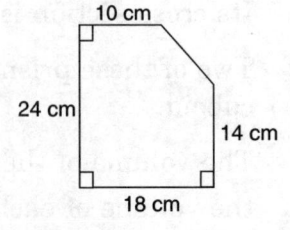

The diagram shows the cross-section of a concrete curb stone which is 1 m long.

Calculate the volume of the curb stone:
a) in cubic centimetres
b) in cubic metres

Solution

The cross-section can be regarded as a rectangle 24 cm by 18 cm with a right-angled triangle cut off one corner. The base of the triangle is $(18 - 10)$ cm and the height is $(24 - 14)$ cm.

a) Area of cross-section $= (24 \times 18) \text{ cm}^2 - \frac{1}{2}(8 \times 10) \text{ cm}^2$
$= (432 - 40) \text{ cm}^2$
$= 392 \text{ cm}^2$

Volume of curb stone $= \text{area of cross-section} \times \text{length}$
$= 392 \times 100 \text{ cm}^3$
$= 39\ 200 \text{ cm}^3$

b) 1 cubic metre = 1 metre × 1 metre × 1 metre
= 100 cm × 100 cm × 100 cm
= 1 000 000 cm³

Hence, volume of curb stone = $\frac{39\,200}{1\,000\,000}$ m³
= 0.0392 m³

Example 3

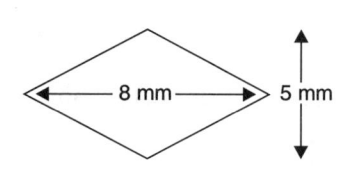

A pendant in the shape of a prism has a cross-section which is a rhombus with diagonals of length 8 mm and 5 mm. The pendant is 3 cm long.

Calculate the volume of the pendant.

Solution

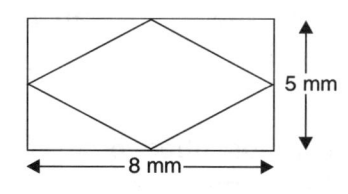

The diagonals of a rhombus are at right angles and so the rhombus can be enclosed in a rectangle 8 mm by 5 mm, as shown in the diagram.

Area of rhombus = $\frac{1}{2}$ (area of rectangle)
= $\frac{1}{2}$(8 × 5) mm²
= 20 mm²

Volume of pendant = area of rhombus × length
= 20 mm² × 30 mm
= 600 mm³

Volume of a cylinder

A cylinder is a prism with a circular cross-section.

The circular cross-section can be regarded as a regular polygon with a very large number of sides.

It follows that we can find the volume of a cylinder by using the formula for the volume of a prism. The area of cross-section is πr^2, so if the height of the cylinder is represented by h, we have

$$\text{volume of a cylinder} = \pi r^2 h$$

This is another formula which you must remember.

Example 1

A cylindrical garden roller has a diameter of 50 cm and a length of 65 cm.

Calculate the volume of the roller.

Solution

The radius of the roller = $\frac{50}{2}$ cm = 25 cm

The volume of the roller = $\pi r^2 h$
$= 3.142 \times (25)^2 \times 65$ cm^3
$= 127\,643.75$ cm^3
$= 128\,000$ cm^3 to 3 significant figures

> since 1 m^3 = 1 000 000 cm^3, the volume could be written as 0.128 m^3

Example 2

A large cylindrical tin of coffee powder has a radius of 8 cm and a height of 20 cm.

a) Calculate the volume of the tin.

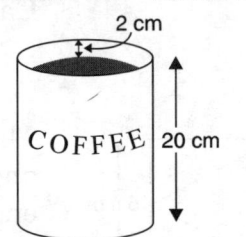

b) On opening the tin, Rachelle finds that there is a gap of depth 2 cm between the top of the tin and the surface of the coffee.

Calculate the volume of coffee in the tin.

c) To make a cup of coffee, 5 cm^2 of coffee powder is required. Calculate the number of cups of coffee that can be made from the tin.

Solution

a) Volume of the cylindrical tin = $\pi r^2 h$
$= 3.142 \times 8^2 \times 20$ cm^3
$= 4021.76$ cm^3
$= 4020$ cm^3 to 3 significant figures

b) Depth of coffee in the tin = $(20 - 2)$ cm = 18 cm
Volume of coffee in the tin = $\pi r^2 h$
$= 3.142 \times 8^2 \times 18$ cm^3
$= 3619.584$ cm^3
$= 3620$ cm^3 to 3 significant figures

c) $\frac{\text{volume of coffee in the tin}}{\text{volume of coffee for 1 cup}} = \frac{3619.584 \text{ cm}^3}{5 \text{ cm}^3} = 723.9168$

Hence, the number of cups of coffee which can be made from the tin = 723.

> we used 3619.584 cm^3 for the volume of the coffee in the tin, not the 'corrected' value obtained as the answer to part b)

Example 3

Some copper wire has a circular cross-section with a diameter of 1.2 mm.

Calculate the volume, in cubic centimetres, of copper in a 150 m length of the wire.

Solution

Radius of the wire = $\frac{1.2 \text{mm}}{2}$ = 0.6 mm = 0.06 cm

Length of the wire = 150 m = 15 000 cm

Volume of copper = $\pi r^2 h$
$= 3.142 \times (0.06)^2 \times 15\,000$ cm^3
$= 169.668$ cm^3
$= 170$ cm^3 to 3 significant figures

Exercise 13 contains questions on prisms and cylinders. You will have to choose the appropriate formulae for each question. Remember that area is measured in *square* units and area formulae contain a length times a length. Volume is measured in *cubic* units and volume formulae contain a product of three lengths.

EXERCISE 13

1. Calculate the volume of:
   a) a prism with cross-sectional area 1.4 cm^2 and length 11 cm
   b) a cylindrical coin with diameter 22 mm and thickness 3 mm

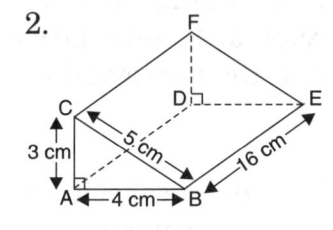

2. The diagram shows a triangular prism ABCDEF.
   Angle CAB = 90°, AB = 4 cm, BC = 5 cm, CA = 3 cm and BE = 16 cm.
   a) Calculate the surface area of the prism.
   b) Calculate the volume of the prism.

3. The diagram represents a cylindrical well. It is 10 m deep and has a diameter of 1.6 m.
   a) (i) Write down the radius of the well.
      (ii) Calculate the volume of the well.
   b) The well is dug through rock. Every cubic metre of this rock has a mass of 2.3 tonnes. Calculate the mass of the rock removed in digging the well.

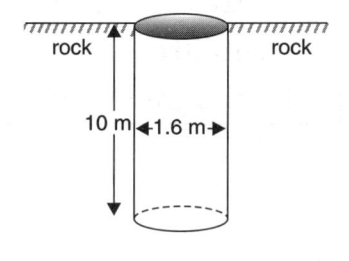

4. The diagram shows the cross-section of some wooden picture frame moulding. It is drawn on a 1 cm grid.
   a) Write down the area of the cross-section.
   b) Calculate the volume of a 3 m length of the moulding.

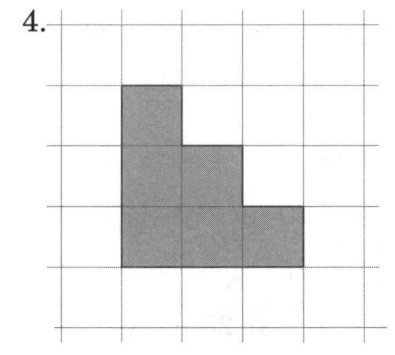

5. A manufacturer is making a batch of 800 plastic knitting needles. Each needle is cylindrical in shape, with a diameter of 6 mm and a length of 32 cm. Calculate the total volume of plastic used in making these needles.

Check your answers at the end of this module.

Capacity

Hollow containers such as cups, bottles, rectangular plastic tubes, etc., can be used to hold liquids or other 'fluids' such as toothpaste. The **capacity** of a container is the volume of liquid or fluid it can hold.

Capacity is sometimes measured in the same units as volume (cubic centimetres, cubic metres, etc.) but, because liquids and fluids can take any shape, the units for measuring them should not, strictly speaking, refer to length.

The most common unit for measuring liquids and fluids is the **litre**, and this is the basic unit for measuring the capacity of a container. For example, petrol (and the capacity of a car's petrol tank) is measured in litres.

Sometimes it is necessary to use a smaller unit. Wine can be measured in centilitres and a dose of medicine may be measured in millilitres.

Because the capacity of a container is measured by the volume of liquid or fluid it can contain, the units of capacity and the units of volume must be related. These relationships are shown in the table below, together with the abbreviations for the units which are commonly used.

Abbreviations	Units of capacity and volume	Units of Capacity
ℓ = litre	1 litre = 1000 cm^3	1 litre = 100 cℓ
cℓ = centilitre	1 centilitre = 10 cm^3	1 litre = 1000 mℓ
mℓ = millilitre	1 millilitre = 1 cm^3	1 cℓ = 10 mℓ

Example 1

A bottle contains $\frac{1}{2}$ ℓ of medicine. John has to take 4 teaspoonfuls every day. How long will the medicine last? (1 teaspoonful = 5 mℓ.)

Solution

$\frac{1}{2}$ ℓ = 500 mℓ

4 teaspoonsful = 20 mℓ

$\frac{500}{20} = 25$

The medicine will last 25 days.

Example 2

A water tank has a rectangular base 60 cm by 90 cm.
The tank is filled at a rate of $4\frac{3}{4}$ ℓ per minute.
How long will it take to fill the tank to a depth of 40 cm?

Solution

$4\frac{3}{4}\,\ell = 4\frac{3}{4} \times 1000 \text{ cm}^3 = 4750 \text{ cm}^3$

Volume of water required $= 60 \times 90 \times 40 \text{ cm}^3$
$= 216\,000 \text{ cm}^3$

Time required $= \frac{216\,000}{4750}$ minutes $= 45.47\ldots$ minutes
$= 45.5$ minutes to 3 significant figures

In the IGCSE examinations, your understanding of capacity may be tested by short questions, or by part of a longer question on mensuration. Here are some short questions for you to try.

EXERCISE 14

1. How many 5 millilitre spoonfuls can be obtained from a bottle that contains 0.3 litres of medicine?

2. A bottle contains 75 cℓ of wine. A wine glass holds 70 mℓ. How many wine glasses can be completely filled from 8 bottles of wine?

3. A cylindrical can of fizzy drink has a radius of 4 cm and a height of 12 cm.

   Calculate, in litres, the amount of drink it will hold.

Check your answers at the end of this module.

C Extending the use of the formulae

There is often more than one stage to solving a mensuration problem and you may have to use more than one of the mensuration formulae. You will have to decide which formulae are needed.

> In IGCSE examinations you are expected to give answers correct to 3 significant figures, unless you are told to use some other degree of accuracy. In order to achieve this degree of accuracy, you should keep at least 4 significant figures in your working until you have the answer. Then you should correct to 3 significant figures. Do not be tempted to correct intermediate results to 3 significant figures, or to use answers to previous parts of the question which have been corrected to 3 figures.

Example 1

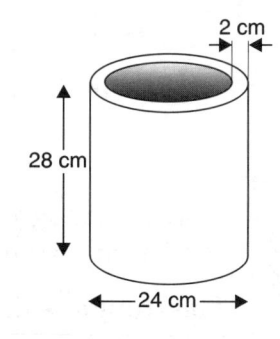

A large plant pot is cylindrical in shape. It is made of clay 2 cm thick and is open at the top.

The external diameter of the pot is 24 cm and its external height is 28 cm.
a) Find the internal diameter and the internal height of the pot.
b) Find the volume of clay used to make the pot.

Solution

a) Internal diameter $= 24 \text{ cm} - 4 \text{ cm} = 20 \text{ cm}$
   Internal height $= 28 \text{ cm} - 2 \text{ cm} = 26 \text{ cm}$

b) The pot can be regarded as a cylinder of diameter 24 cm and height 28 cm, with a cylinder of diameter 20 cm and height 26 cm taken away.

Remembering that in $\pi r^2 h$ we must use *radius,* not diameter,

$$\begin{aligned}\text{volume of clay} &= (\pi \times (12)^2 \times 28 - \pi \times (10)^2 \times 26) \text{ cm}^3 \\ &= (4032\pi - 2600\pi) \text{ cm}^3 \\ &= 1432\pi \text{ cm}^3 \\ &= 1432 \times 3.142 \text{ cm}^3 \\ &= 4499.344 \text{ cm}^3 \\ &= 4500 \text{ cm}^3 \text{ to 3 significant figures}\end{aligned}$$

Example 2

A water tank is in the shape of a cuboid, measuring 150 cm by 100 cm by 80 cm.
a) How many litres of water would the tank hold when full?
b) The tank is initially empty and water flows into it from a pipe. The cross-sectional area of the pipe is 2.1 cm² and the water flows along the pipe at a rate of 35 cm/s.
Find the time taken to fill the tank. Give your answer in hours and minutes, correct to the nearest minute.

Solution

a) Capacity of the tank = $150 \times 100 \times 80$ cm³
$\qquad\qquad\qquad\qquad\;\; = 1\,200\,000$ cm³

$1 \ell = 1000$ cm³ so capacity of the tank $= 1200\, \ell$.

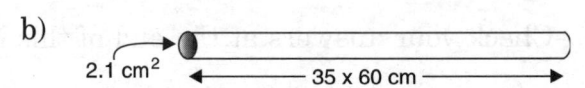

b)

2.1 cm²   ←— 35 × 60 cm —→

Volume of water which flows out of the pipe in 1 minute
= volume of a prism with cross-section 2.1 cm² and length
  (35×60) cm
= $2.1 \times (35 \times 60)$ cm³
= 4410 cm³

Time to fill the tank = $\dfrac{1\,200\,000}{4410}$ minutes = 272.1 minutes

$\qquad\qquad\qquad\qquad$ = 4 hours 32 minutes to the nearest minute

Example 3

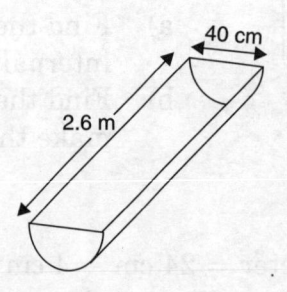

A water trough is 2.6 m long and has a semicircular cross-section with diameter 40 cm.

The trough is fixed so that its rectangular top is horizontal.
a) Calculate the capacity of the trough in litres.
b) If the trough is full of water, calculate the area of the surface which is in contact with the water.

Solution

a) Radius of the cross-section = $\frac{40}{2}$ cm = 20 cm

Area of cross-section = $\frac{1}{2} \times \pi(20)^2$ cm^2

$= 200\pi$ cm^2

Length of the trough = 2.6 m = 260 cm

Capacity of the trough = $200\pi \times 260$ cm^3
$= 52\,000 \times 3.142$ cm^3
$= 163\,384$ cm^3
$= 163.384$ ℓ
$= 163$ ℓ to 3 significant figures

b) Area in contact with the water
= two semicircular ends
   + half curved surface area of cylinder

$= \frac{1}{2}\pi(20)^2 + \frac{1}{2}\pi(20)^2 + \frac{1}{2} \times (2\pi \times 20 \times 260)$ cm^2

$= (200\pi + 200\pi + 5200\pi)$ cm^2
$= 5600\pi$ cm^2
$= 5600 \times 3.142$ cm^2
$= 17\,595.2$ cm^2
$= 17\,600$ cm^2 to 3 significant figures

Example 4

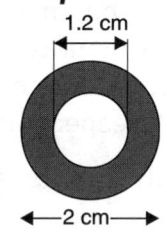

The diagram shows the cross-section of a brass washer. The washer is 3 mm thick.
a) Calculate the volume of the washer.
b) Calculate the surface area of the washer.

Solution

a) Outer radius = 1 cm and inner radius = 0.6 cm

Area of cross-section = $(\pi \times 1^2 - \pi \times (0.6)^2)$ cm^2
$= (\pi - 0.36\pi)$ cm^2
$= 0.64\pi$ cm^2

Thickness of washer = 3 mm = 0.3 cm

Volume of washer = $(0.64\pi \times 0.3)$ cm^3
$= 0.603264$ cm^3
$= 0.603$ cm^3 to 3 significant figures

b) From part a), area of top face of washer = 0.64π cm^2
and area of bottom face of washer = 0.64π cm^2

Inner curved surface area = $2\pi \times (0.6) \times (0.3)$ cm^2
$= 0.36\pi$ cm^2

Outer curved surface area = $2\pi \times (1) \times (0.3)$ cm^2
$= 0.60\pi$ cm^2

Total surface area = $(0.64 + 0.64 + 0.36 + 0.60)\pi$ cm^2
$= 2.24 \times 3.142$ cm^2
$= 7.03808$ cm^2
$= 7.04$ cm^2 to 3 significant figures

Mass and density

The **mass** of an object is the amount of matter in it. The greater the mass of the object, the more difficult it is to get it moving – or, more accurately, the more difficult it is to accelerate it.

In everyday life, the mass of an object is often confused with its weight. The **weight** of an object is the force with which the earth (or other heavenly body on which the object is located) attracts the object. A man on the moon weighs much less than he does on earth (because the moon's attraction is much less than the earth's) but his *mass* is exactly the same on the moon as it is on the earth.

Although it may seem strange to talk about the mass of a bag of potatoes or the mass of a new-born baby (when you usually say 'weight'), it is technically correct, and in our work we will use the correct word.

One standard unit of mass is the **gram** – you probably know that the postage charge for a letter or small package depends on its mass in grams. A gram is quite a small unit and so you will find that, in many everday situations, a **kilogram** is used. A kilogram is 1000 grams. For more massive objects, a unit of a **tonne** may be used. A tonne is 1000 kilograms.

In examination questions (as well as in everyday life), you will find that abbreviations are used – 'g' for grams, 'kg' for kilograms, and 't' for tonnes.

Density is a measure of compactness of a substance and is its mass per unit volume. Density is usually expressed in grams per cubic centimetre or kilograms per cubic metre. These units are abbreviated to g/cm^3 and kg/m^3, respectively.

If a solid is of uniform material, its density is found by dividing its mass by its volume.

You need to remember:

$$\text{density} = \frac{\text{mass}}{\text{volume}}$$

Example 1

Gold has a density of 19.3 g/cm^3.
Find the mass, in kilograms, of a rectangular block of gold which has dimensions 9 cm by 5 cm by 3 cm.

Solution

Volume of rectangular block = length × breadth × height
　　so volume of gold = 9 × 5 × 3 cm^3
　　　　　　　　　　 = 135 cm^3

1 cm^3 of gold has a mass of 19.3 grams.

Hence, mass of the block of gold = 135 × 19.3 g
　　　　　　　　　　　　　　　　 = 2605.5 g
　　　　　　　　　　　　　　　　 = 2.6055 kg
　　　　　　　　　　　　　　　　 = 2.61 kg to 3 significant figures

Example 2

A cylinder of lead has a radius of 3.2 cm and a height of 5.7 cm.
The mass of the cylinder is 2.08 kg.
Using this information, calculate the density of lead.

Solution

Volume of cylinder = $\pi r^2 h$

Volume of lead = $3.142 \times (3.2)^2 \times 5.7$ cm^3
 = $183.39\ldots$ cm^3

Mass of lead = 2.08 kg = 2080 g

Density of lead = $\dfrac{\text{mass}}{\text{volume}}$

= $\dfrac{2080 \text{ g}}{183.39 \text{ cm}^3}$

= $11.34\ldots$ g/cm^3

= 11.3 g/cm^3 to 3 significant figures

Changing the subject

Formulae such as

curved surface area of a cylinder = $2\pi rh$
volume of a cuboid = length × breadth × height
volume of a prism = area of cross-section × height
volume of a cylinder = $\pi r^2 h$

are useful when you want to calculate the subject of the formula and you know the values of all the other quantities in the formula.

If you want to calculate the value of a quantity which is not the subject of the formula, you will either have to change the subject of the formula or you will have to form an equation and solve it.
You will remember that we faced the same problem when we were dealing with plane figures in Unit 1 Section C.

Example 1

Find a formula for calculating the radius of a cylinder when you know its height and its volume.

Solution

The formula for the volume of a sphere is $V = \pi r^2 h$

Dividing both sides by πh, we get $\dfrac{V}{\pi h} = r^2$

Taking the positive square root of both sides $\sqrt{\dfrac{V}{\pi h}} = r$

The formula required is $r = \sqrt{\dfrac{V}{\pi h}}$

Example 2

A rectangular tank has a base 35 cm by 25 cm.
What is the depth of water in the tank if it contains 14 ℓ of water?

Solution

14 ℓ = 14 000 cm³

Let the depth of the water be d cm.

Using volume of cuboid = length × breadth × height,
$$14\,000 = 35 \times 25 \times d$$
$$14\,000 = 875d$$
$$d = \frac{14\,000}{875} = 16$$

The depth of water in the tank = 16 cm.

Example 3

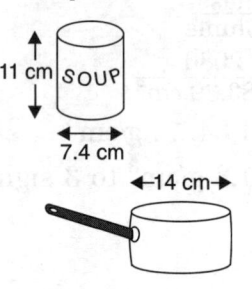

A cylindrical can of soup has a diameter of 7.4 cm and a height of 11 cm.

The soup is poured into a cylindrical pan which has a diameter of 14 cm.

Calculate the depth of soup in the pan, to the nearest millimetre.

Solution

Radius of the can = 3.7 cm and radius of the pan = 7 cm.

Let the depth of soup in the pan be d cm.

Volume of soup in the can = $\pi r^2 h$
$$= \pi \times (3.7)^2 \times 11 \text{ cm}^3$$

Volume of soup in the pan = $\pi \times (7)^2 \times d$ cm³

Hence $\pi \times (7)^2 \times d = \pi \times (3.7)^2 \times 11$
$$d = \frac{\pi \times (3.7)^2 \times 11}{\pi \times (7)^2} = 3.073\ldots$$

Depth of soup in the pan = 3.1 cm to the nearest millimetre.

> We did not substitute any particular value for π because it 'cancels out'. You would obtain the same final answer no matter what value of π you use.

Example 4

A cylindrical tank has an internal radius of 16 cm. It stands upright on its base on horizontal ground and contains water to a depth of 25 cm. A cubical block of metal with side 12 cm is immersed in the water. Calculate the increase in depth of the water.

Solution

Volume of the metal block = length × breadth × height
$$= 12 \text{ cm} \times 12 \text{ cm} \times 12 \text{ cm}$$
$$= 1728 \text{ cm}^3$$

Let the increase in depth of the water be d cm.

The extra volume of the tank occupied $= \pi \times (16)^2 \times d$ cm^3.
This extra volume must be the same as the volume of the block.
Hence $\pi \times (16)^2 \times d = 1728$

$$d = \frac{1728}{3.142 \times (16)^2} = 2.148$$

The increase in depth of the water $= 2.15$ cm to 3 significant figures.

> We do not seem to have used the original depth (25 cm) of the water. However, in order to obtain the answer to the problem we need to know that the original depth is sufficient for the block of metal to be *completely* immersed in the water.

The questions in Exercise 15 are not easy. You will need to think hard and work carefully.

EXERCISE 15

1. A sheet of window glass is $1\frac{1}{2}$ m long, 1 m wide and 2 mm thick.
   a) Calculate the volume of the sheet in cubic centimetres.
   b) The density of glass is 2.8 g/cm^3. Calculate the mass of the sheet.

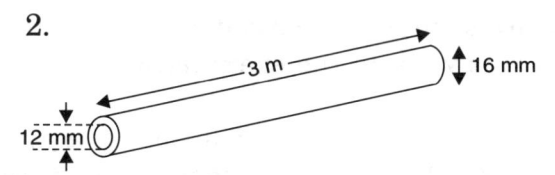

2. A cylindrical metal tube is 3 m long. The external diameter of the tube is 16 mm and the internal diameter is 12 mm.

   Calculate the volume of the metal making up the tube.

3. A solid cylinder has a radius of 5 cm and a surface area of 270 cm^2.
   a) Calculate the area of the curved surface of the cylinder.
   b) Calculate the height of the cylinder, to the nearest millimetre.

4. A hose pipe has an internal circular cross-section with radius 0.8 cm. Water flows through the pipe at a rate of 30 cm/s.
   a) Calculate the volume of water which is delivered by the pipe in 1 minute.
   b) The water from the hose pipe flows into a rectangular tank with a base 25 cm by 20 cm. By how much does the water level in the tank rise in 1 minute? Give your answer correct to 2 significant figures.

5. A cylindrical bottle has an internal radius of 6 cm. It contains apple juice to a depth of 12 cm. The apple juice is poured into cylindrical beakers. The radius of each beaker is 3 cm.
   How many of these beakers can be filled to a depth of 7 cm with apple juice from the bottle?

Check your answers at the end of this module.

If you have any wrong answers, examine the solutions I have given. You may need to do some revision of the work on mensuration.

If you are following the CORE syllabus, you have almost completed Unit 2. Well done! You should now turn to the 'Summary'.

If you are following the EXTENDED syllabus, you need to study the material in Section D.

D Surface area and volume of pyramids, cones and spheres

For the IGCSE EXTENDED syllabus examinations, you are expected to use mensuration formulae for pyramids, cones and spheres as well as the formulae you have already met for cuboids, prisms and cylinders.

You are not expected to be able to *prove* these formulae (in fact, to prove them properly you need to know some advanced mathematics, such as the calculus). You are not even required to *remember* the formulae for pyramids, cones and spheres – if a formula is needed to solve a problem, the examiner will give the formula in the question paper.

In some cases, we can indicate how a formula could be obtained or we can show a connection between two formulae. The 'notes' written below should be regarded as optional reading – you will not be tested on this work in the examination.

Surface area and volume of a pyramid

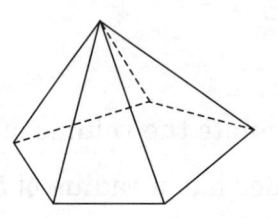

The base of a pyramid is a polygon and all the other faces are triangles.

To find the surface area of a pyramid, you find the area of the base and the area of each of the triangles, and add them together.

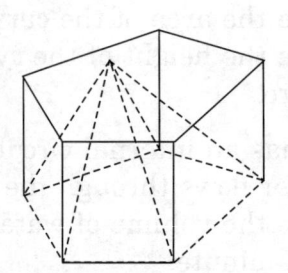

Alternatively, you could find the area of the net of the pyramid.

The volume of a pyramid is one third the base area × perpendicular height.

This means that it is exactly one third of the volume of the prism which has the same base and the same perpendicular height.

The formula we shall use is:

$$\text{volume of a pyramid} = \tfrac{1}{3} \times \text{area of base} \times \text{perpendicular height}$$

Note: A cube can be cut up into three congruent square-based pyramids. The diagrams below show how this is done.

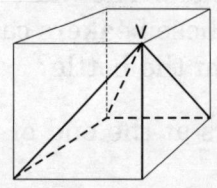

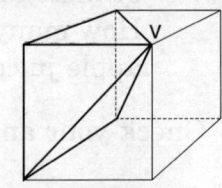

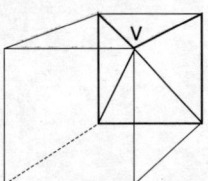

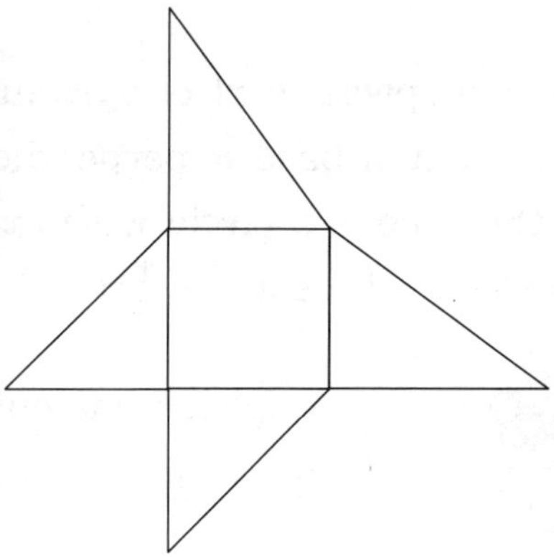

Surface area and volume of a cone

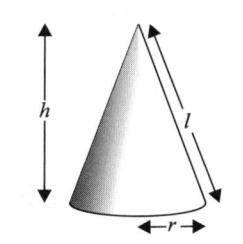

You could make three pyramids for yourself and see how they fit together to make a cube.

Here is a net for one pyramid (the other two are exactly the same as this one).

The cube is a prism, so its volume is area of base × perpendicular height. Hence, the volume of each pyramid is $\frac{1}{3}$ × area of base × perpendicular height.

A **cone** is a special sort of pyramid – one which has a circular base. We shall consider only **right cones** (cones which are upright) these cones have their axis of symmetry perpendicular to the base.

The radius of the base is usually denoted by r and the **perpendicular** height is denoted by h.

The **slant height** is denoted by l (see the diagram).

The formulae you will need are:

curved surface area of a cone = $\pi r l$, where l = slant height

volume of a cone = $\frac{1}{3}\pi r^2 h$, where h = perpendicular height

Note 1:

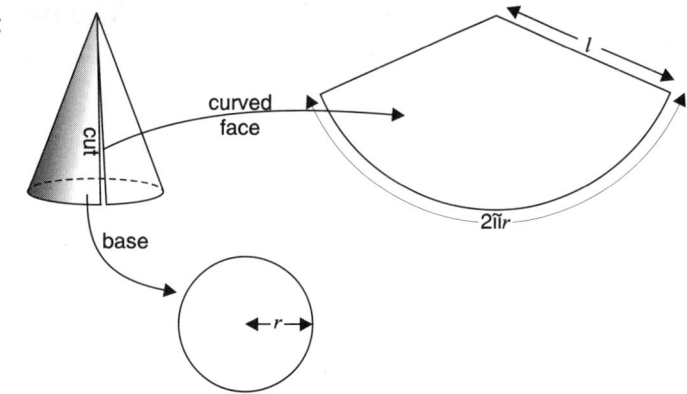

A cone has two faces – the flat circular base and the curved face.

The base has an area of πr^2.

The curved face can be cut along a line from the vertex to a point on the circumference of the base, and flattened out to make a sector of a circle.

The radius of this sector is l and its arc has a length of $2\pi r$ (it was originally the circumference of the base).

The total circumference of a circle of radius l is $2\pi l$, so the sector is a fraction $\frac{2\pi r}{2\pi l}$ of the whole circle. Hence, its area is $\frac{2\pi r}{2\pi l} \times \pi l^2 = \pi r l$.

Thus, the curved surface area of the cone = $\pi r l$.

Surface area and volume of a sphere

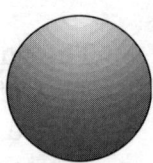

Note 2:
Since a cone is a special sort of pyramid, we can use
volume = $\frac{1}{3}$ × area of base × perpendicular height.
The base of the cone is a circle with area πr^2.
Hence, the volume of a cone is $\frac{1}{3}\pi r^2 h$.

A sphere has only one face, and it is curved.
The radius of the sphere is denoted by r.

The formulae you will need are:

$$\text{surface area of a sphere} = 4\pi r^2$$

$$\text{volume of a sphere} = \frac{4}{3}\pi r^3$$

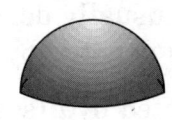

A **hemisphere** is half a sphere.
It has one flat face (a circle with an area of πr^2) and one curved face (with an area of $2\pi r^2$).
The total surface area of a hemisphere = $3\pi r^2$.
The volume of a hemisphere = $\frac{1}{2} \times \frac{4}{3}\pi r^3 = \frac{2}{3}\pi r^3$.

Note:

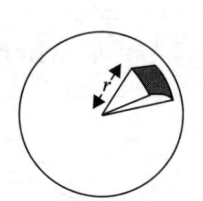

A solid sphere can be regarded as a collection of pyramids with vertex at the centre of the sphere and base on the surface of the sphere.

The perpendicular height of each pyramid is r, the radius of the sphere.

The volume of each pyramid
= $\frac{1}{3}$ × area of base × perpendicular height.

Hence, the volume of the sphere = $\frac{1}{3}$ × total area of the base × r
= $\frac{1}{3}$ × surface area of the sphere × r
= $\frac{1}{3}$ × $4\pi r^2$ × r
= $\frac{4}{3}\pi r^3$

Example 1

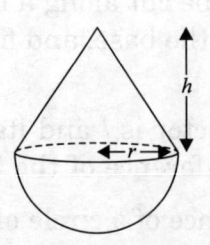

The diagram shows a solid which consists of a hemisphere of radius r, attached to a cone of base radius r and height h.
a) Calculate the volume of the solid when $r = 8$ cm and $h = 10$ cm.
b) Calculate the value of h when the volume of the solid is 3000 cm³ and $r = 10$ cm. Give your answer correct to the nearest millimetre.

Solution

a) The volume of the solid V is given by the formula
$V = \frac{2}{3}\pi r^3 + \frac{1}{3}\pi r^2 h$.

When $r = 8$ cm and $h = 10$ cm,
$$V = \frac{2}{3}\pi(8)^3 + \frac{1}{3}\pi(8)^2(10) \text{ cm}^3$$
$$= \frac{2\pi}{3}(8)^2(8+5) \text{ cm}^3$$
$$= \frac{128\pi}{3}(13) \text{ cm}^3$$
$$= 1742.762\ldots \text{ cm}^3$$
$$= 1740 \text{ cm}^3 \text{ to 3 significant figures}$$

b) When $V = 3000$ cm^3 and $r = 10$ cm,
$$\tfrac{2}{3}\pi(10)^3 + \tfrac{1}{3}\pi(10)^2 h = 3000$$
$$\tfrac{1}{3}\pi(10)^2 h = 3000 - \tfrac{2}{3}\pi(10)^3$$
$$h = \left(\frac{3 \times 3000}{\pi(10)^2} - 2(10)\right) \text{ cm}$$
$$h = \left(\frac{90}{\pi} - 20\right) \text{ cm}$$
$$h = 8.644\ldots \text{ cm}$$
$$h = 8.6 \text{ cm to nearest millimetre}$$

Example 2

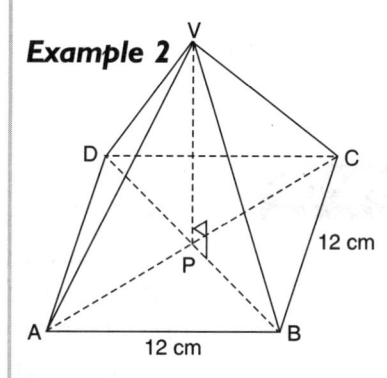

The diagram represents a pyramid on a square base of side 12 cm. The diagonals AC and BD meet at P. The perpendicular height, PV, is 9.5 cm.
a) Calculate the volume of the pyramid.
b) How many planes of symmetry has the pyramid?

Solution

a) Volume of pyramid $= \frac{1}{3} \times$ area of base $\times$ perpendicular height
$$= \tfrac{1}{3} \times 144 \times 9.5 \text{ cm}^3$$
$$= 456 \text{ cm}^3$$

b) The pyramid has 4 planes of symmetry.
(The square base has 4 lines of symmetry. Each plane of symmetry of the pyramid passes through one of these lines and through the vertex V.)

Example 3

A solid sphere has a radius of 2.4 cm.
a) Calculate the volume of the sphere.
b) The density of gold is 19.32 g/cm^3. Assuming that the solid sphere is all gold, calculate its mass to the nearest gram.

c) In fact, the mass of the sphere is only 852 g because it contains some lead as well as gold. The density of lead is 11.34 g/cm³. Calculate the volume of lead in the sphere.

Solution

a) Volume of sphere $= \frac{4}{3}\pi r^3$

$= \frac{4}{3} \times 3.142 \times (2.4)^3$ cm³

$= 57.913344$ cm³

$= 57.9$ cm³ to 3 significant figures

b) 1 cm³ of gold has a mass of 19.32 g.
Mass of gold sphere $= 57.913344 \times 19.32$ g
$= 1118.885\ldots$ g
$= 1119$ grams to the nearest gram.

c) Every cubic centimetre of lead reduces the mass of the sphere by $(19.32 - 11.34)$ g, that is 7.98 g.

In fact, the mass of the sphere has been reduced by $(1118.885 - 852)$ g, that is 266.885 g.

Volume of lead in the sphere $= \frac{266.885}{7.98}$ cm³

$= 33.44\ldots$ cm³

$= 33.4$ cm³ to 3 significant figures

Are you ready to solve problems on pyramids, cones and spheres? Here are a few for you to try.

EXERCISE 16

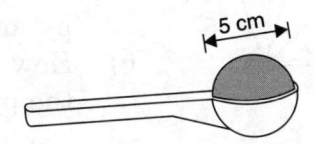

1. An ice-cream scoop forms ice-cream in the shape of a sphere of diameter 5 cm.
   a) Taking π to be 3.14, find the volume of a sphere of ice-cream. Give your answer correct to one decimal place.
   b) How many spheres of ice-cream can be made from 1 ℓ of ice-cream?

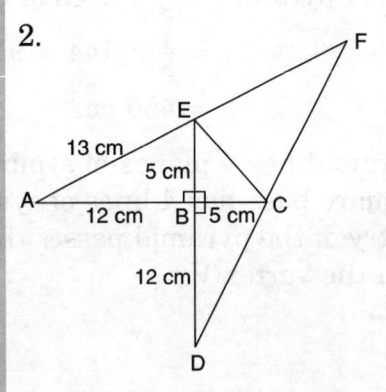

2. The diagram shows the net of a solid. The straight lines ABC and DBE are perpendicular.
   EB = BC = 5 cm, AB = BD = 12 cm, AE = 13 cm.
   a) Of what special type is this solid?
   b) Write down the length of EF.
   c) Calculate the volume of the solid.

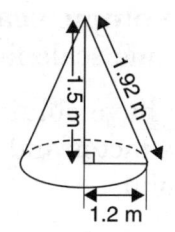

3. The diagram shows a heap of sand in the shape of a cone. The radius of the base of the cone is 1.2 m, its perpendicular height is 1.5 m. And its slant height is 1.92 m.
   a) Calculate the volume of the sand.
   b) Calculate the curved surface area of the cone.

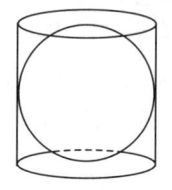

4. A glass sphere has a radius of 5 cm.
   a) Calculate the volume of the sphere.
   b) The sphere is tightly packed in a cylindrical gift box, touching the curved surface of the box and both the top and the bottom.
      (i) Calculate the total surface area of the interior of the box.
      (ii) Express the volume of the sphere as a fraction of the capacity of the box.

Check your answers at the end of this module.

Surface areas and volumes of similar solids

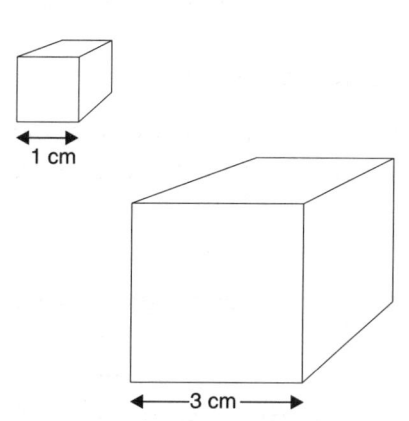

Look at these two cubes.

The edges of the larger cube are *3 times* as long as the edges of the smaller cube. This means that the **linear scale factor** is 3.

What is the surface area scale factor?

What is the volume scale factor?

The larger cube has 6 square faces, each with an area of 9 cm². Its total surface area is 54 cm².

The smaller cube has 6 square faces, each with an area of 1 cm². Its total surface area is 6 cm².

Hence, the surface area of the larger cube is *9 times* the surface area of the smaller cube. The **surface area scale factor** is 9.

You will notice that $9 = 3^2$, so we can say
   the surface area scale factor = (linear scale factor)².

This should remind you of a result we had for similar plane shapes in Unit 1.

The volume of the larger cube is $3 \times 3 \times 3$ cm³, that is 27 cm³.
The volume of the smaller cube is $1 \times 1 \times 1$ cm³, that is 1 cm³.

This shows that the **volume scale factor** = 27, and since $27 = 3^3$, we can write the volume scale factor = (linear scale factor)^3.

The relationships we have obtained between the linear scale factor, the surface area scale factor and the volume scale factor are true for any pair of similar solids.

Consider, for example, a circular cone with radius r, perpendicular height h and slant height l.
Its surface area is $\pi rl + \pi r^2$ and its volume is $\frac{1}{3}\pi r^2 h$.

Suppose a similar cone has radius kr, perpendicular height kh, and slant height kl. Then the linear scale factor = k.

$$\begin{aligned}\text{The surface area of the second cone} &= \pi(kr)(kl) + \pi(kr)^2 \\ &= k^2\pi rl + k^2\pi r^2 \\ &= k^2(\pi rl + \pi r^2)\end{aligned}$$

Hence, the surface area scale factor = k^2.

$$\begin{aligned}\text{Also, the volume of the second cone} &= \tfrac{1}{3}\pi(kr)^2(kh) \\ &= \tfrac{1}{3}\pi k^2 r^2 kh \\ &= k^3\left(\tfrac{1}{3}\pi r^2 h\right)\end{aligned}$$

Hence, the volume scale factor = k^3 = (linear scale factor)^3.

If you prove the results for prisms, cylinders, pyramids and spheres, you will soon realise that they depend on the facts that the formulae for surface areas contain two lengths multiplied together, and the formulae for volumes contain three lengths multiplied together.

You need not worry about these proofs, but you must remember the results:

> For a pair of similar solids,
> surface area scale factor = (linear scale factor)^2
> volume scale factor = (linear scale factor)^3

Example 1

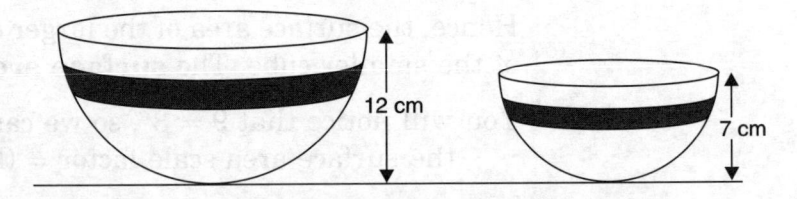

The bowls shown in the diagram are similar. The capacity of the smaller bowl is 300 mℓ. Calculate the capacity of the larger bowl.

Solution

The linear scale factor $= \frac{12}{7}$.

Capacity is the *volume* of liquid a container will hold.

$$\begin{aligned}\text{Volume scale factor} &= (\text{linear scale factor})^3 \\ &= \left(\frac{12}{7}\right)^3 \\ &= \frac{1728}{343}\end{aligned}$$

$$\begin{aligned}\text{Volume of larger bowl} &= 300 \text{ m}\ell \times \frac{1728}{343} \\ &= 1511.37\ldots \text{ m}\ell \\ &= 1510 \text{ m}\ell \text{ to 3 significant figures}\end{aligned}$$

Example 2

A 'Regular Size' tube of toothpaste, which is 12 cm long, contains 50 mℓ of toothpaste when full. A similar 'Giant Size' tube of toothpaste is an enlargement of the 'Regular Size' tube, with a scale factor of 2.
a) What is the length of the 'Giant Size' tube?
b) How many millilitres of toothpaste will the 'Giant Size' tube hold when full?

Solution

a) 'Enlargement with scale factor of 2' means that the linear scale factor is 2.

$$\begin{aligned}\text{Hence, length of the 'Giant Size' tube} &= 12 \text{ cm} \times 2 \\ &= 24 \text{ cm}\end{aligned}$$

b) $$\begin{aligned}\text{Volume scale factor} &= (\text{linear scale factor})^3 \\ &= 2^3 \\ &= 8\end{aligned}$$

$$\begin{aligned}\text{Hence, amount of toothpaste in the 'Giant Size' tube} &= 50 \text{ m}\ell \times 8 \\ &= 400 \text{ m}\ell\end{aligned}$$

Example 3

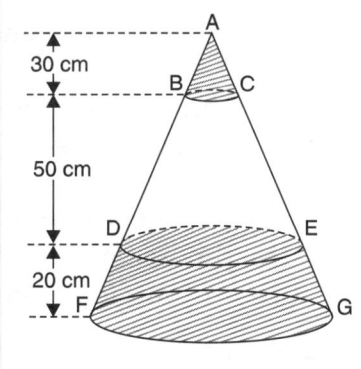

The diagram represents a traffic cone. The vertical heights of the three sections are 30 cm, 50 cm and 20 cm, as shown.

The top and bottom sections, shown shaded, are painted red.
a) What is the ratio of the three slant heights, AB : AD : AF?
b) What is the ratio of the areas of the curved surfaces of the cones ABC, ADE and AFG?
c) What fraction of the curved surface of the traffic cone is painted red? Give your answer in its lowest terms.

Solution

a) The vertical heights of the cones ABC, ADE and AFG are 30 cm, 80 cm and 100 cm. These cones are similar to one another. Hence, the ratio of the slant heights is 30 : 80 : 100 (using the linear scale factors). This simplifies to 3 : 8 : 10.

b) The surface area scale factors are the squares of the linear scale factors. Hence, the ratio of the curved surface areas of the cones ABC, ADE, AFG = 9 : 64 : 100.

c) From part b), the surface area of cone ABC = 9 units
the surface area of cone ADE = 64 units
the surface area of cone AFG = 100 units

The red sections are ABC (9 units) and DEFG (100 − 64 units = 36 units). These add up to 45 units. This means that $\frac{45}{100}$ of the whole cone (AFG) is painted red. That is, $\frac{9}{20}$ of the curved surface of the traffic cone is painted red.

Many learners find it difficult to grasp the idea of different scale factors for length, area and volume. See whether you can answer the following questions.

EXERCISE 17

1. Two similar jugs have heights 10 cm and 16 cm.
   The smaller jug holds 500 mℓ.
   Calculate the capacity of the larger jug.

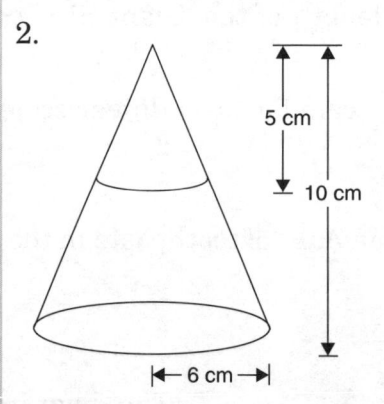

2. A solid cone has a radius of 6 cm and a vertical height of 10 cm.
   a) Calculate the volume of the cone.
   b) The cone is cut into two pieces by a plane parallel to the base. The smaller piece is a cone with vertical height 5 cm.
      (i) What is the base radius of this smaller piece?
      (ii) What is the volume of this smaller piece?

3. Two similar solids have surface areas of 200 cm^2 and 450 cm^2. The larger solid is regarded as an enlargement of the smaller solid.
   a) Find: (i) the surface area scale factor
             (ii) the linear scale factor
             (iii) the volume scale factor
   b) The volume of the larger solid is 1350 cm^3.
      Calculate the volume of the smaller solid.

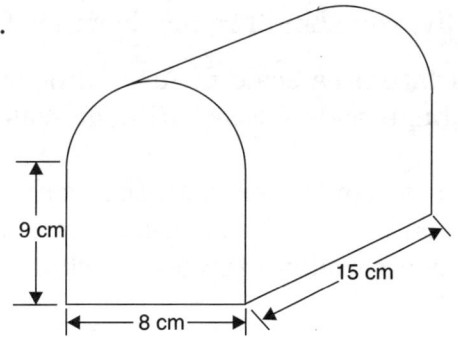

4. A loaf of bread has the shape of a prism, 15 cm long.

Its cross-section consists of a rectangle, 8 cm by 9 cm, with a semicircular top.

a) Calculate the volume of the loaf.

b) For publicity, the manufacturer of the bread decides to have a large balloon made in a similar shape to the loaf. The enlargement is to have a scale factor of 50.
   (i) Calculate the volume of the balloon.
   (ii) Calculate the area of the material, in square metres, needed to make the balloon.

If you found this exercise difficult, don't worry too much. This topic is only a minor one for the IGCSE examinations. Nevertheless, you should check the answers at the end of this module and if any of them are wrong, you should read the solutions carefully and try to understand how to solve the problems.

Summary

You should remember the following formulae for calculating surface areas and volumes of solids:

- surface area of a cuboid $= 2LB + 2BH + 2HL$
- curved surface area of a cylinder $= 2\pi rh$ add πr^2 for the top and another πr^2 for the bottom
- volume of a cuboid $=$ length $\times$ breadth $\times$ height
- volume of a prism $=$ area of cross-section $\times$ height
- volume of a cylinder $= \pi r^2 h$.

Remember also that capacity is volume of liquid a container can hold, usually measured in litres or millilitres.

I explained the difference between mass and weight. The units of mass most commonly used are the kilogram, gram or tonne.

The density of a substance $= \frac{\text{mass}}{\text{volume}}$.

Once again in this unit you have had practise when changing the subject of the formula.

Those of you studying the EXTENDED syllabus went on to learn about the surface area and volume of pyramids, cones and spheres. In an exam you will be reminded of the following formulae if you need to use them in a calculation:

- volume of a pyramid $= \frac{1}{3} \times$ area of base $\times$ perpendicular height
- curved surface area of a cone $= \pi r l$, where $l =$ slant height
- volume of a cone $= \frac{1}{3}\pi r^2 h$, where $h =$ perpendicular height
- surface area of a sphere $= 4\pi r^2$
- volume of a sphere $= \frac{4}{3}\pi r^3$.

Finally, you should remember that for a pair of similar solids:

- surface area scale factor = (linear scale factor)^2
- volume scale factor = (linear scale factor)^3.

You have now covered all the work on mensuration but, before you move on to Unit 3, you should attempt the 'Check your progress'. This is a revision exercise which covers a variety of ideas from Unit 2.

Check your progress

1. A glass has a capacity of 120 mℓ.
   a) How many full glasses of orange juice can be obtained from a 1 ℓ carton?
   b) How much orange juice would be left over?

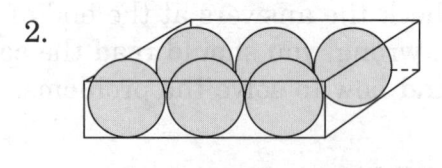

2. The radius of a tennis ball is 3.2 cm. The balls are sold in packs of six.

   The balls are tightly packed in a rectangular box, as shown in the diagrams.
   a) Write down the internal length, breadth and height of the box.
   b) Calculate the volume of the box.
   c) Calculate the surface area of the inside of the box.

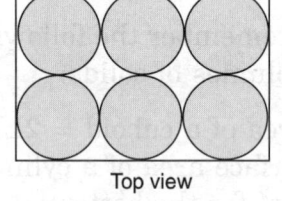

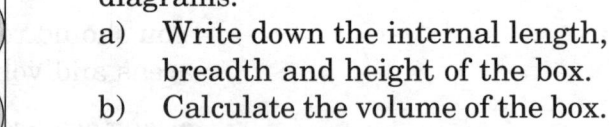

3. The diagram shows a litter bin in the shape of a cylinder. It has a base but no top.

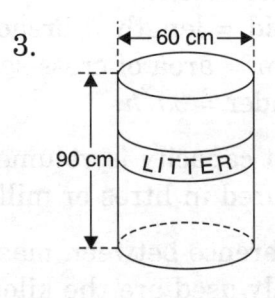

   The diameter of the bin is 60 cm and its height is 90 cm.
   a) Write down the radius of the litter bin.
   b) Calculate the volume of the litter bin.
   c) (i) Calculate the area of metal used to make the curved surface of the litter bin.
   (ii) Calculate the area of metal used to make the base of the bin.
   (iii) Show clearly that the total area of metal used to make the litter bin, correct to 2 significant figures, is 2.0 m^2.

d)

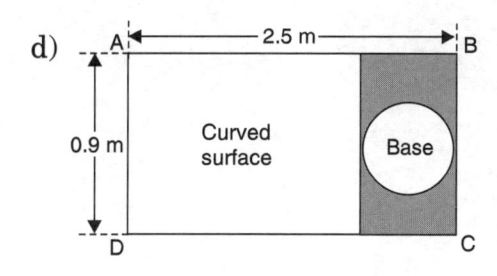

The curved surface and base of the litter bin were cut out from a rectangular piece of metal, ABCD, 2.5 m long and 0.9 m wide, as shown. The shaded part of the rectangle is wasted.

(i) Calculate the area of ABCD, in square metres.
(ii) Using the information given in part c) (iii), calculate the percentage of the metal wasted in making the litter bin.

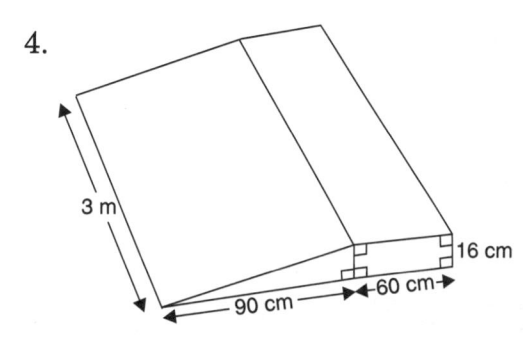

4. A concrete ramp is to be built outside the door of a building to allow access for people with wheelchairs.

The ramp consists of a triangular prism and a cuboid, with dimensions as shown in the diagram.

Calculate, in cubic metres, the volume of:
a) the triangular prism
b) the cuboid
c) the complete ramp

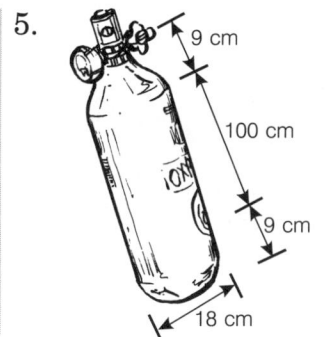

5. Oxygen is stored in containers which have the shape of a cylinder with a hemisphere at each end, as shown in the diagram.
a) Calculate the total surface area of the outside of the cylinder.
b) Calculate the total volume of the cylinder.

Check your answers at the end of this module.

Unit 3
Right-angled Triangles and Trigonometry

The word **trigonometry** means **triangle measurement**. The first book to use trigonometry in its title was written in the late 16th century by a German mathematician-astronomer called Bartholomew Pitiscus. However, if we regard trigonometry as the geometry of similar triangles, its roots can be traced back to the Greek mathematician-astronomers Aristarchus, Hipparchus, Menelaus and Ptolemy in the period 300 BC to AD 150.

In Module 4, you learnt how to solve problems by scale drawing. Trigonometry to some extent replaces scale drawing by methods of calculation which give more accurate results. In the form of **triangle measurement** it is used in surveying, navigation, engineering and astronomy, but you will find trigonometrical expressions and calculations are also used in subjects such as electricity, magnetism and electronics.

This unit is divided into four sections:

Section	Title	Time
A	Pythagoras's theorem	2 hours
B	Trigonometry – the tangent ratio	3 hours
C	The sine and cosine ratios	$2\frac{1}{2}$ hours
D	Using trigonometry to solve problems	$2\frac{1}{2}$ hours

By the end of this unit, you should be able to:

- use Pythagoras's theorem to calculate the length of the third side of a right-angled triangle when the lengths of the other two sides are known
- determine whether a triangle is right-angled when the lengths of its sides are known
- use trigonometry to calculate lengths and angles in right-angled triangles
- use a scientific calculator for trigonometrical calculations.

A Pythagoras's theorem

One of the most important results in geometry is known as Pythagoras's theorem. This is a result which is true for all right-angled triangles. You already know that the three *angles* of a triangle add up to 180°. Pythagoras's theorem tells us something about the three *sides* of a right-angled triangle.

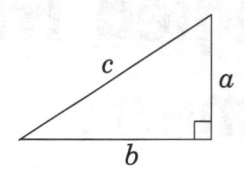

If the length of the side opposite the right-angle is c and the lengths of the other two sides are a and b, then:

$$c^2 = a^2 + b^2$$

Pythagoras was a Greek mathematician and philosopher who lived in the 6th century BC. It is doubtful whether Pythagoras himself proved the theorem. It was probably first proved by one of his contemporaries. Hundreds of years earlier, the Egyptians knew that a triangle with sides of 3, 4 and 5 units is right-angled. They used this fact in their surveying and construction of pyramids, etc. In India, the Hindus not only knew about the 3, 4, 5 triangle but were aware of other right-angled triangles, for example those with sides 5, 12 and 13 units or 8, 15 and 17 units.

As far as we know, the Egyptians and Hindus did not look for the connection between the lengths of the sides of a right-angled triangle. To discover this connection required the sort of mathematical investigation which fascinated the Greeks. They regarded Pythagoras's theorem as a statement about *areas*.

In a right-angled triangle, the side opposite the right angle is called the **hypotenuse**. The hypotenuse is always the longest side in the triangle.

Pythagoras's theorem can be stated as follows:

> In a right-angled triangle, the area of the square on the hypotenuse is equal to the sum of the areas of the squares on the other two sides.

This Greek stamp shows a right-angled triangle with squares drawn on its three sides.

You can see that the area of the square on the hypotenuse is 25 square units and the areas of the squares on the other two sides are 16 square units and 9 square units.

The fact that $25 = 16 + 9$ confirms Pythagoras's theorem for this right-angled triangle.

We cannot be sure which was the first proof of Pythagoras's theorem. There are now hundreds of known proofs and demonstrations. (There are 256 of them in a book called 'The Pythagorean Proposition' by Elisha Scott Loomis.)

Here is one of the best-known demonstrations:

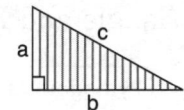

Let the sides of the right-angled triangle be a, b and c, where c is the hypotenuse.

Consider two identical squares, each with sides of length $(a + b)$. From each square, cut off four right-angled triangles with sides a, b, c, as shown in the diagrams below.

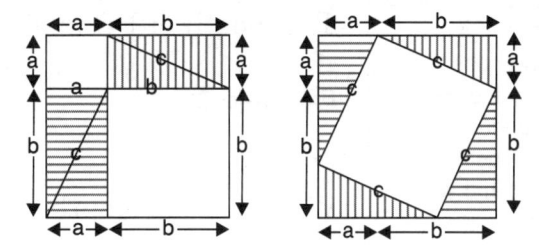

Since the original squares had the same area, and equal amounts have been cut off them, the parts which are left must have the same area.

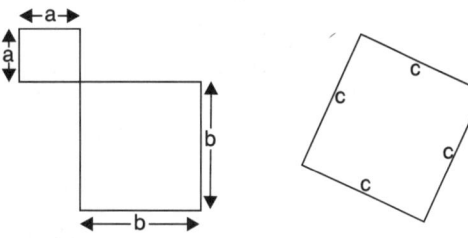

Hence, $c^2 = a^2 + b^2$.

This is the statement of Pythagoras's theorem in algebraic form. You will realise that the word 'square' can be used to mean a geometrical shape or it can be used in its algebraic sense of 'multiply a number by itself'. The Greeks thought of the geometrical square because they did not develop algebra to any great extent. When we use Pythagoras's theorem, we almost always think of 'square' in its algebraic sense.

In effect, Pythagoras's theorem means that, if you know the lengths of any two sides of a right-angled triangle, you can always calculate the length of the third side. This is extremely useful because right-angled figures occur so often in theoretical and practical problems.

Example 1

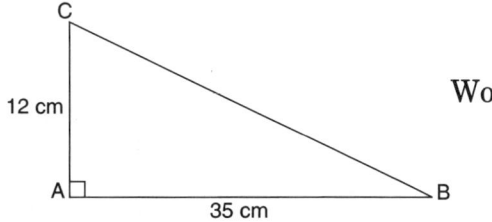

Work out the length of BC.

You could do the whole of this calculation on your calculator, as follows:

[1] [2] [SHIFT] [x^2] [+] [3]
[5] [SHIFT] [x^2] [=] [√]

However, when you are answering examination questions, it is essential that you show the working. As you press the keys, write down the calculator display at every step. You will get marks for correct working even if your final answer is wrong.

Solution

ABC is a right-angled triangle and BC is the hypotenuse.
Let BC $= x$ cm
Then $x^2 = 12^2 + 35^2$ — Pythagoras's theorem
$= 144 + 1225$
$= 1369$
$x = \sqrt{1369} = 37$

The length of BC is 37 cm.

Example 2

Work out the length of QR.

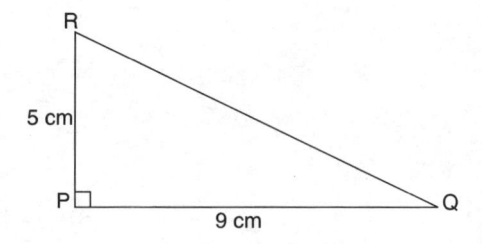

Solution

PQR is a right-angled triangle and QR is the hypotenuse.
Let QR $= x$ cm
Then $x^2 = 5^2 + 9^2$ — Pythagoras's theorem
$= 25 + 81$
$= 106$
$x = \sqrt{106} = 10.2956\ldots$ — calculator display

The length of QR is 10.3 cm to 3 significant figures.

Example 3

Work out the length of YZ.

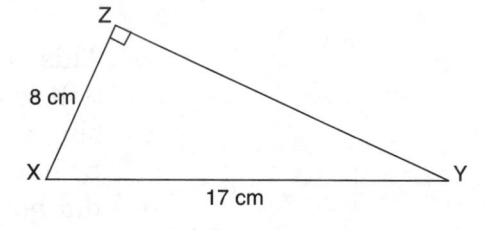

Solution

XYZ is a right-angled triangle with hypotenuse XY = 17 cm.
Let YZ $= x$ cm.
Then $17^2 = 8^2 + x^2$ — Pythagoras's theorem
$17^2 - 8^2 = x^2$
$x^2 = 289 - 64$
$= 225$
$x = \sqrt{225} = 15$

The length of YZ is 15 cm.

When applying Pythagoras's theorem, you should always start off with (hypotenuse)$^2 = \ldots$
When you have the lengths of all three sides of the triangle, check that the longest side is opposite the right angle.

Example 4

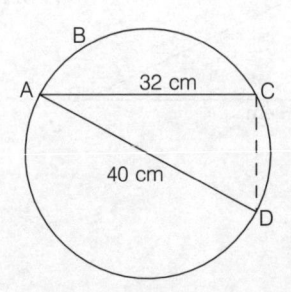

Kurt has a circular log of wood, of diameter 40 cm.

He saws it in half along AD. He then cuts off another piece, whose cross-section is ABC, so that AC = 32 cm.

a) Give a reason why angle ACD = 90°.
b) Calculate the straight line distance from C to D.

Solution

a) AD is a diameter of the circle, so angle ACD is 90° because it is an angle in a semicircle.

b) ACD is a right-angled triangle with hypotenuse AD = 40 cm.
Let CD = x cm

$$\text{Then } 40^2 = 32^2 + x^2 \quad \text{[Pythagoras's theorem]}$$
$$40^2 - 32^2 = x^2$$
$$x^2 = 1600 - 1024$$
$$= 576$$
$$x = \sqrt{576} = 24$$

The straight line distance from C to D is 24 cm.

Example 5

The point A has coordinates (−2, 1) and the point B has coordinates (3, 4). Calculate the length of the line segment AB.

Solution

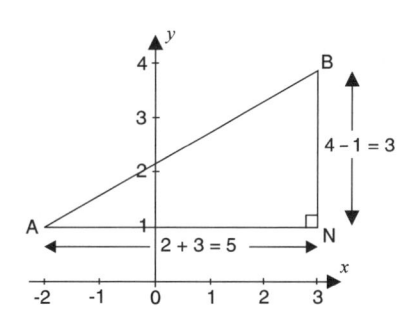

To solve this problem, we use the horizontal and vertical distances, AN and BN, between A and B. The lines AN and BN, together with the line AB, form a right-angled triangle.

AN = 2 + 3 = 5 and BN = 4 − 1 = 3.
ABN is a right-angled triangle with hypotenuse AB.

Let AB = x units
$$x^2 = 5^2 + 3^2 \quad \text{[Pythagoras's theorem]}$$
$$= 25 + 9$$
$$= 34$$
$$x = \sqrt{34} = 5.8305\ldots$$

The length of AB = 5.83 to 3 significant figures.

Example 6

The base for a garage is a rectangle RSTU with side RS = 4.25 m and side ST = 5.30 m.
Calculate the length of the diagonal RT, correct to 3 significant figures.

Solution

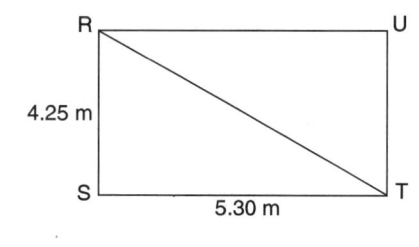

Angle RST = 90° (angle of a rectangle).
RST is a right-angled triangle with hypotenuse RT.
Let RT = x m
$$x^2 = (4.25)^2 + (5.30)^2 \quad \text{[Pythagoras's theorem]}$$
$$= 18.0625 + 28.09$$
$$= 46.1525$$
$$x = \sqrt{46.1525} = 6.79356\ldots$$

The length of the diagonal RT is 6.79 m to 3 significant figures.

Example 7

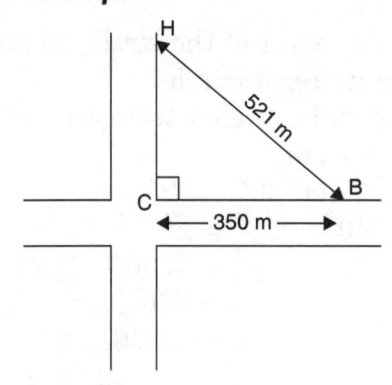

Mohamed takes a short cut from his home (H) to the bus stop (B) along a footpath HB.

How much further would it be for Mohamed to walk to the bus stop by going from H to the corner (C) and then from C to B?

Give your answer in metres.

Solution

If Mohamed takes the alternative route, the distance he walks is HC + CB.

HCB is a right-angled triangle with hypotenuse HB = 521 m.
Let HC = x m.
$$(512)^2 = (350)^2 + x^2 \quad \text{[Pythagoras's theorem]}$$
$$x^2 = (521)^2 - (350)^2$$
$$= 271\,441 - 122\,500$$
$$= 148\,941$$
$$x = \sqrt{148\,941} = 385.9287$$

Length of alternative route = (385.9287 + 350) m
= 735.9287 m

Extra distance walked = (735.9287 − 521) m
= 214.9287 m
= 215 m to 3 significant figures

Have you understood Pythagoras's theorem and how it can be used? For success in solving problems, you will have to recognise right angles in various situations. For example:

- the angles of a square or rectangle
- the angle between horizontal and vertical
- the angle between a north-south line and an east-west line
- the angle in a semicircle
- the angle between a tangent and the radius at its point of contact
- the angle between the diagonals of a rhombus or kite.

When you have obtained an answer, you should make sure that it is sensible. Always check that the longest side is opposite the right angle, and that it is shorter than the other two sides added together (otherwise a triangle cannot be formed).

Common mistakes are to add squares when they should be subtracted, and to forget to take the square root as the final step.

Test your understanding of Pythagoras's theorem by answering the following questions.

EXERCISE 18

1. For each of the triangles shown below, work out the length of the lettered side.

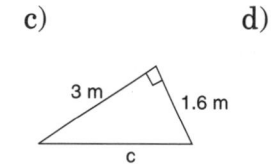

   a) 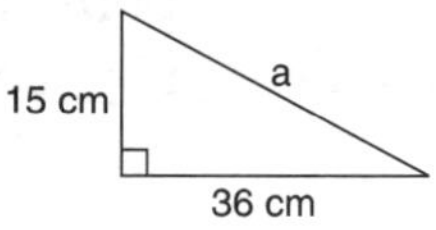 [triangle: 15 cm vertical, 36 cm horizontal, hypotenuse a]

   b) [triangle: 25 cm hypotenuse, 7 cm base, vertical b] 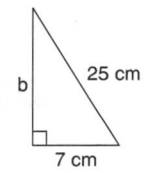

   c) [triangle: 3 m and 1.6 m legs with right angle, hypotenuse c]

   d) [triangle: 60 mm and 65 mm, d]

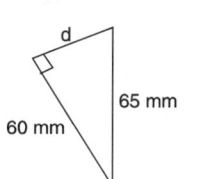

2. 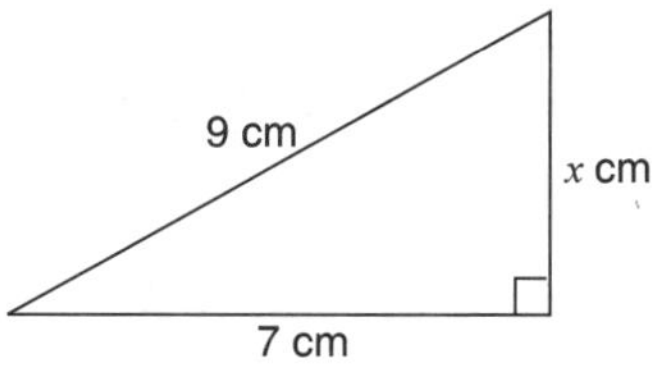 [triangle: 9 cm hypotenuse, 7 cm base, vertical x cm] Work out the value of x.

3. 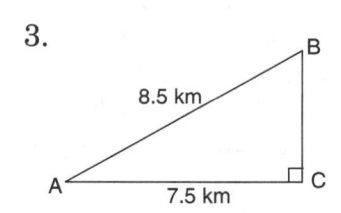 [triangle ABC: AB = 8.5 km, AC = 7.5 km, right angle at C]

   In the diagram, AB = 8.5 km, AC = 7.5 km and angle ACB = 90°.

   Calculate the length of BC.

4. 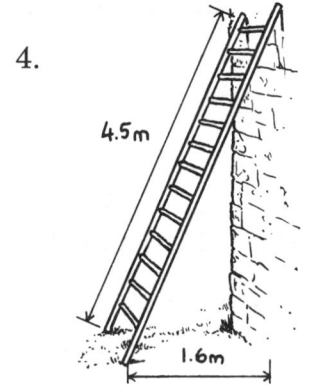 [ladder against wall: 4.5 m ladder, 1.6 m from wall]

   A ladder is standing on horizontal ground and rests against a vertical wall. The ladder is 4.5 m long and its foot is 1.6 m from the wall.

   Calculate how far up the wall the ladder will reach. Give your answer correct to 3 significant figures.

5. Grisley is 90 km due north of Ford.
   Highton is 64 km due east of Ford.
   Calculate the distance of Highton from Grisley, giving your answer correct to 3 significant figures.

6. 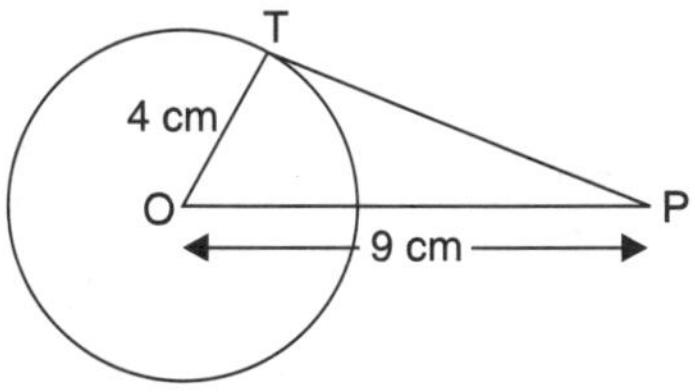 [circle with centre O, radius 4 cm to T, OP = 9 cm, tangent PT]

   A circle has centre O and radius 4 cm. A point P is 9 cm from O. A tangent is drawn from P to the circle and its point of contact is T.

   a) Give a reason why angle OTP = 90°.

   b) Calculate the length of PT.

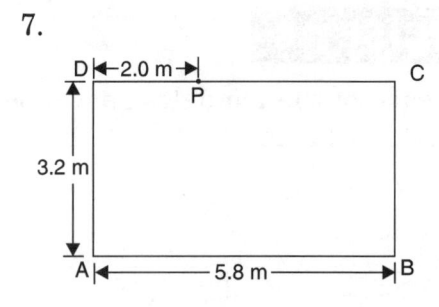

7. In the diagram, the rectangle ABCD represents the floor of a room. P represents a power point at floor level.

Calculate the distance of the power point from:
a) the corner A of the room
b) the corner B of the room

You shouldn't have found these problems too difficult to solve if you remembered to identify the hypotenuse of the relevant right-angled triangle, and used Pythagoras's theorem in the form (hypotenuse)^2 = ...

Check your answers at the end of this module.

Further applications of Pythagoras's theorem

It is important to remember that Pythagoras's theorem is concerned with *right-angled triangles*. Sometimes, to solve a problem, you may have to create right-angled triangles by drawing extra lines which are perpendicular to lines already in the diagram. This is particularly useful when you are dealing with isosceles triangles. Sometimes, the right-angled triangles are in a 3-dimensional situation and, in a 2-dimensional diagram, they may not look right-angled.

Example 1

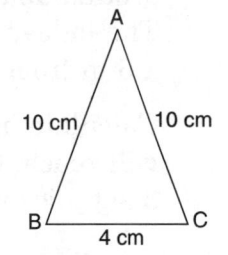

A triangle ABC has AB = 10 cm, BC = 4 cm and CA = 10 cm.

Calculate the area of the triangle.

Solution

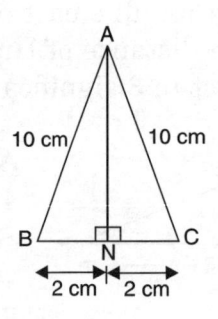

This triangle is isosceles. (The sides AB and CA are the same length.)

To find its area, we need to know its perpendicular height.

In this diagram, AN is the perpendicular from A to BC. Because of the symmetry of the isosceles triangle, BN = NC = 2 cm.

Triangle ABN is right-angled with hypotenuse AB = 10 cm.

Let AN = h cm. Then $10^2 = 2^2 + h^2$ [Pythagoras's theorem]
$$h^2 = 10^2 - 2^2$$
$$= 100 - 4$$
$$= 96$$
$$h = \sqrt{96} = 9.79795\ldots$$ [calculator display]

Area of triangle ABC $= \frac{1}{2} \times$ base $\times$ perpendicular height
$= \frac{1}{2} \times 4 \times 9.79795 \ldots$ cm^2
$= 19.5959 \ldots$ cm^2
$= 19.6$ cm^2 to 3 significant figures

Example 2

A rectangular box has a base with internal dimensions 21 cm by 28 cm, and an internal height of 12 cm.
Calculate the length of the longest straight thin rod which will fit:
a) on the base of the box
b) in the box

Solution

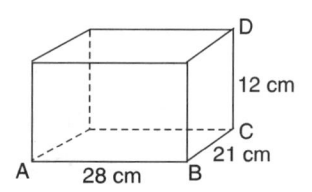

This diagram represents the rectangular box. The box contains many right angles, but most of them do not look like right angles in this diagram!

a) The longest rod which will fit on the base of the box could be placed along AC.

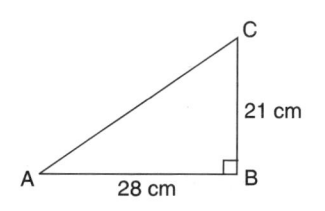

You have to recognise that angle ABC $= 90°$.
Triangle ABC is right-angled with hypotenuse AC.

$(AC)^2 = (28)^2 + (21)^2$ [Pythagoras's theorem]
$= 784 + 441$
$= 1225$
$AC = \sqrt{1225} = 35$ cm

The length of the longest rod which will fit on the base of the box is 35 cm.

b) The longest rod which will fit in the box could be placed along the diagonal AD of the box.

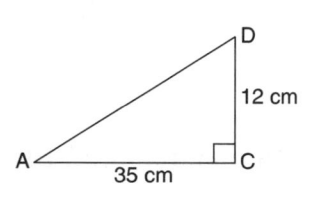

You have to recognise that angle ACD $= 90°$. Triangle ACD is right-angled with hypotenuse AD.

$(AD)^2 = (35)^2 + (12)^2$
$= 1225 + 144$
$= 1369$
$AD = \sqrt{1369} = 37$ cm   [Pythagoras's theorem]

The length of the longest rod which will fit in the box is 37 cm.

Example 3

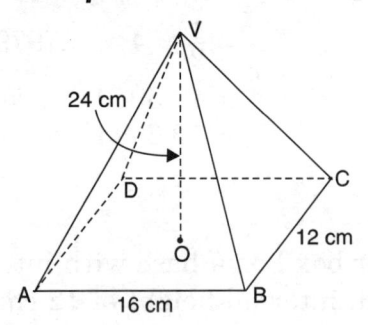

The diagram represents a pyramid with a rectangular horizontal base ABCD. AB = 16 cm and BC = 12 cm.

The vertex V of the pyramid is vertically above the centre O of the base and VO = 24 cm.
a) Calculate the length of the diagonal AC of the base.
b) Calculate the length of AO.
c) Calculate the length of the sloping edge AV of the pyramid.

Solution

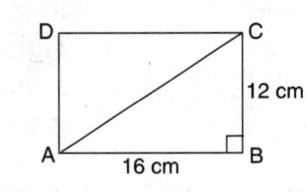

a) The base of the pyramid is a rectangle and so triangle ABC is right-angled. Its hypotenuse is AC.
$(AC)^2 = (16)^2 + (12)^2$ [Pythagoras's theorem]
$ = 256 + 144$
$ = 400$
$AC = \sqrt{400}$
$AC = 20$ cm

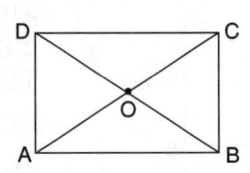

b) O is the centre of the base, so AO = OC.
Hence $AO = \frac{AC}{2}$
$AO = 10$ cm

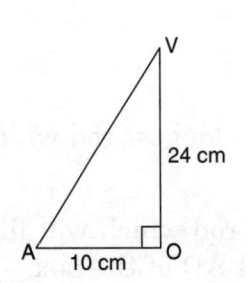

c) The base is horizontal and V is vertically above O, so angle AOV = 90°.
Triangle AOV is right-angled with hypotenuse AV.
$(AV)^2 = (10)^2 + (24)^2$ [Pythagoras's theorem]
$ = 100 + 576$
$ = 676$
$AV = \sqrt{676}$
$AV = 26$ cm

Example 4

A rhombus PQRS has diagonal PR = 13 cm and diagonal QS = 9.4 cm.
Calculate the lengths of the sides of the rhombus.

Solution

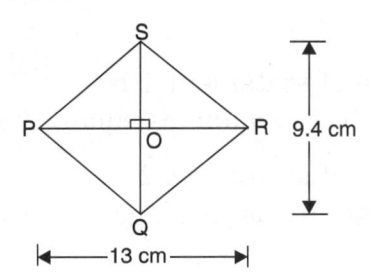

The diagonals of a rhombus bisect each other at right angles.

Hence, if O is the point where the diagonals cross, triangle POS is right-angled, with PO = 6.5 cm and OS = 4.7 cm.

PS is the hypotenuse of triangle POS, and so
$(PS)^2 = (6.5)^2 + (4.7)^2$ [Pythagoras's theorem]
$= 42.25 + 22.09$
$= 64.34$
$PS = \sqrt{64.34} = 8.02122\ldots$ [calculator display]

The sides of the rhombus are all equal. Each side has a length of 8.02 cm to 3 significant figures.

Example 5

A cone has a radius of 5 cm and a perpendicular height of 8 cm. Calculate the area of the curved surface of the cone.

Solution

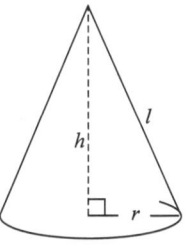

The formula for the curved surface area of a cone is $\pi r l$, where l is the slant height. The relationship between the radius (r), the perpendicular height (h) and the slant height (l) is given by Pythagoras's theorem.

$$l^2 = h^2 + r^2$$
$$\text{or } l = \sqrt{h^2 + r^2}$$

For the given cone, $l = \sqrt{8^2 + 5^2}$
$= \sqrt{64 + 25}$
$= \sqrt{89}$
$= 9.43398\ldots$

Curved surface area $= \pi r l$
$= 3.142 \times 5 \times 9.43398\ldots \text{ cm}^2$
$= 148.2078\ldots \text{ cm}^2$
$= 148 \text{ cm}^2$ to 3 significant figures

Checking for right angles

Many mathematical statements have the simple form:
'If ..., then ...'

Here are some examples:
1. If a triangle has two sides equal, then it has two angles equal.
2. If a quadrilateral has all its sides equal, then its diagonals are at right angles.
3. If a triangle has a right angle, then the square of its longest side is equal to the sum of the squares of the other two sides.

These statements can be 'turned round' to give what is called its **converse**.

The converse of statement 1 is:
If a triangle has two angles equal, then it has two sides equal.

The converse of statement 2 is:
If a quadrilateral has its diagonals at right angles, then its sides are equal.

The fact that a statement is true does not ensure that its converse is true. Statement 1 is true and the converse of statement 1 is also true. However, although statement 2 is true, its converse is false. You can see this from the diagrams below.

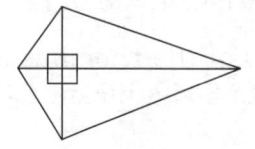

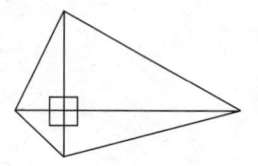

You will have recognised that statement 3 is Pythagoras's theorem and that it is true. The converse of Pythagoras's theorem is:

> If the square of the longest side of a triangle is equal to the sum of the squares of the other two sides, then the triangle has a right angle.

It can be proved that this converse is true, and this gives us a way of testing whether a triangle is right-angled when we know the lengths of its three sides.

Example 1

Is a triangle with sides 5.6 cm, 4.5 cm, 3.1 cm a right-angled triangle?

Solution

The longest side is 5.6 cm and $(5.6)^2 = 31.36$.

$$\text{The sum of the squares of the other two sides} = (4.5)^2 + (3.1)^2$$
$$= 20.25 + 9.61$$
$$= 29.86$$

$(5.6)^2 \neq (4.5)^2 + (3.1)^2$ so the triangle is not right-angled.

Example 2

Is the triangle with sides 7.2 cm, 9 cm, 5.4 cm a right-angled triangle?

Solution

The longest side is 9 cm and $9^2 = 81$.

$$\text{The sum of the squares on the other two sides} = (7.2)^2 + (5.4)^2$$
$$= 51.84 + 29.16$$
$$= 81$$

$9^2 = (7.2)^2 + (5.4)^2$ so the triangle is right-angled.

Example 3

Investigate whether the triangle with sides 11.3 cm, 7.9 cm, 15 cm is acute-angled, right-angled or obtuse-angled.

Solution

This example will show that we can say more than just 'the triangle is right-angled' or 'the triangle is not right-angled'.

The type of triangle depends on the size of the largest angle.
If the largest angle is less than 90°, the triangle is acute-angled.
If the largest angle is 90°, the triangle is right-angled.
If the largest angle is more than 90°, the triangle is obtuse-angled.

The longest side of the given triangle is 15 cm, and $(15)^2 = 225$.

The sum of the squares of the other two sides $= (11.3)^2 + (9.1)^2$
$$= 127.69 + 82.81$$
$$= 210.5$$

$(15)^2 \neq (11.3)^2 + (9.1)^2$ so the triangle is not right-angled.

$(15)^2$ is greater $(11.3)^2 + (9.1)^2$ so the longest side is longer than it would be in a right-angled triangle with shortest sides 11.3 cm and 9.1 cm.

This means that the largest angle (which is opposite the longest side) is more than 90°. In other words, the triangle is obtuse-angled.

To finish this section on Pythagoras's theorem and its converse, here are a few questions for you to answer. You may find them rather more difficult than the questions in the last exercise but, if you understood the examples, you should be able to do them.

EXERCISE 19

1. Each side of an equilateral triangle is 10 cm long.
    a) Calculate the perpendicular height of the triangle.
    b) Calculate the area of the triangle.

2. Each edge of a cube is 20 cm long.
    a) Calculate the length of a diagonal of a face of the cube.
    b) Calculate the length of a diagonal of the cube (the distance between a vertex and the opposite vertex).

3. A circle has centre C and radius 25 cm.
   A chord AB of the circle is 48 cm long.
   Calculate the perpendicular distance from C to AB.

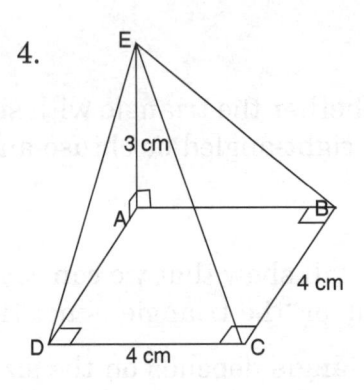

4. The diagram shows a pyramid with a horizontal square base ABCD. The vertex E is vertically above the corner A. AB = BC = CD = DA = 4 cm and AE = 3 cm.

Calculate the length of:
a) EB
b) AC
c) EC

5. Find, by calculation, which of the following triangles are right-angled.

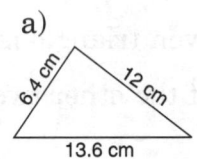

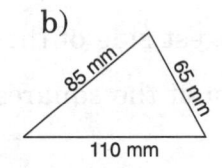

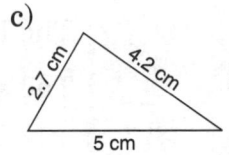

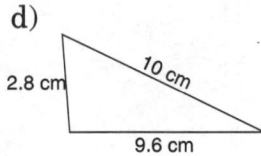

a) 6.4 cm, 12 cm, 13.6 cm

b) 85 mm, 65 mm, 110 mm

c) 2.7 cm, 4.2 cm, 5 cm

d) 2.8 cm, 10 cm, 9.6 cm

6. A triangle has sides of length 53 m, 82 m and 65 m. Investigate whether the triangle is acute-angled, right-angled or obtuse-angled.

Check your answers at the end of this module.

B Trigonometry – the tangent ratio

Look at this problem:

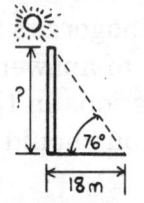

A vertical TV mast casts a shadow 18 m long on horizontal ground when the sun's altitude is 76°. How high is the mast?

To solve this problem using methods we have considered so far, you would have to use scale drawing. You can't use Pythagoras's theorem because Pythagoras's theorem is concerned with sides of right-angled triangles and in the problem you have only been told the length of one side. In this section, you will learn how to use and calculate angles and sides in right-angled triangles. This will enable you to calculate answers to problems (like the one above) without having to do scale drawings. In this way you will obtain more accurate answers.

Naming the sides of a right-angled triangle

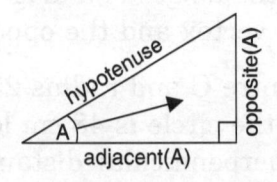

As you already know, the side opposite the right angle is called the **hypotenuse**.

In calculations, we are usually interested in (or concentrating on) one of the acute angles, either because we know its size or we want to know its size. In this program, this angle is labelled A.

When we are working with the angle A, the sides of the triangle (other than the hypotenuse) are named according to their position relative to A.

One side is named **opposite**(A) because it is opposite to the angle A. The other side is named **adjacent**(A) because it is adjacent to (next to) the angle A. The names 'opposite(A)' and 'adjacent(A)' are usually shortened to 'opp(A)' and 'adj(A)'.

> the hypotenuse is also next to angle A so you should always name it first

Examples

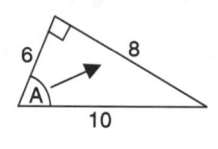

For this triangle:
$$\text{hypotenuse} = 10$$
$$\text{opp}(A) = 8$$
$$\text{adj}(A) = 6$$

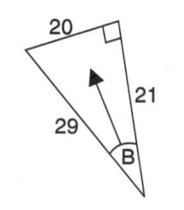

For this triangle:
$$\text{hypotenuse} = 29$$
$$\text{opp}(B) = 20$$
$$\text{adj}(B) = 21$$

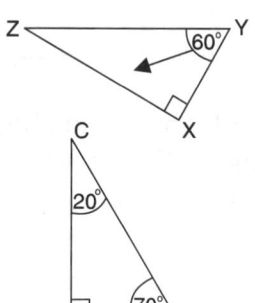

For this triangle:
$$\text{hypotenuse} = YZ$$
$$\text{opp}(60°) = XZ$$
$$\text{adj}(60°) = XY$$

For this triangle:
$$\text{hypotenuse} = BC$$
$$\text{opp}(70°) = AC \quad \text{opp}(20°) = AB$$
$$\text{adj}(70°) = AB \quad \text{adj}(20°) = AC$$

Notice that BC only has one 'name' because it is the hypotenuse. Side AB has two 'names' depending on which of the two angles, 70° and 20°, we are working with. Similarly for side AC.

EXERCISE 20

1. For each of these triangles, write down the length of the hypotenuse and the values of opp(A) and adj(A).

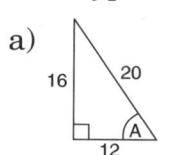

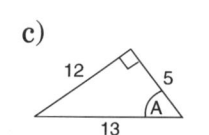

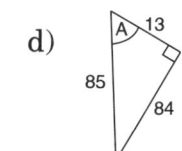

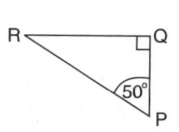

   a)   b)   c)   d)

2. For this triangle, name the hypotenuse and the sides opp(50°) and adj(50°).

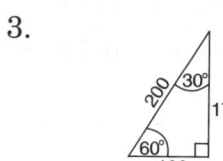

3. For this triangle, write down the length of the hypotenuse and the values of opp(60°), adj(60°), opp(30°) and adj(30°).

Check your answers at the end of this module.

The tangent ratio

Look at these four right-angled triangles. Each of them has an angle of 76°. What else do you notice about them?

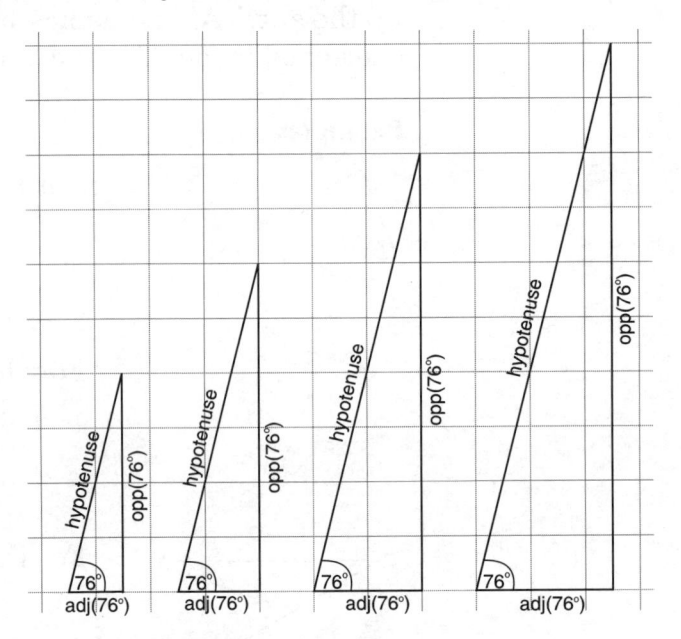

- The triangles are all the same shape.
- They are mathematically similar.
- They are all different sizes. The sides opp(76°) are all different and the sides adj(76°) are all different.

Did you notice anything else?

Did you notice that for every one of these triangles the side opp(76°) is four times as long as the adj(76°)? (It is easier to see when the triangles are drawn on squared paper.)

The fact that the ratio $\frac{\text{opp}(76°)}{\text{adj}(76°)}$ has the same value, 4, for every one of the triangles (no matter how large or small it is) is very important. It is a consequence of the fact that the ratio of corresponding sides of similar triangles is constant.

Of course, if you consider a set of right-angled triangles, each with an angle of (say) 35° the ratio $\frac{\text{opp}(35°)}{\text{adj}(35°)}$ will not be the same as $\frac{\text{opp}(76°)}{\text{adj}(76°)}$, but it will be the same for each of the triangles.

Here are the measurements for these triangles.

Triangle	opp(35°)	adj(35°)	$\frac{\text{opp}(35°)}{\text{adj}(35°)}$
a	1.4 cm	2 cm	0.70
b	2.8 cm	4 cm	0.70
c	2.4 cm	3.45 cm	0.70
d	3.9 cm	5.6 cm	0.70

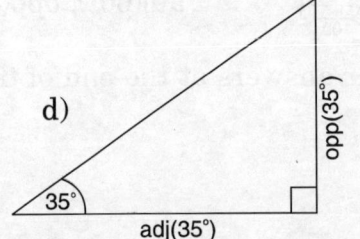

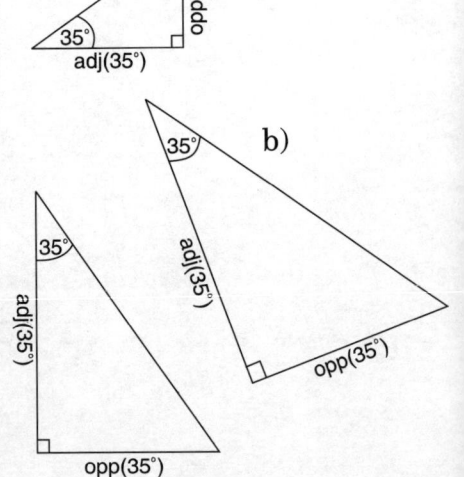

In general, for any set of similar right-angled triangles, the ratio $\frac{\text{opp(angle)}}{\text{adj(angle)}}$ depends on the size of the angle but not on the size of the triangle. This ratio has a special name. It is called the **tangent** of the angle (because it has some connection with the tangent to a circle).

This name is usually shortened to **tan** and we write:

$$\tan(\text{angle}) = \frac{\text{opp(angle)}}{\text{adj(angle)}}$$

where, in place of 'angle' we write the size of the angle or the name of the angle.

For example, $\tan(76°) = \frac{\text{opp}(76°)}{\text{adj}(76°)} = 4$ and $\tan(35°) = \frac{\text{opp}(35°)}{\text{adj}(35°)} = 0.7$.

We could find the value of tan(angle) for any acute angle by drawing and measuring a right-angled triangle which contains that acute angle. To obtain a reasonably accurate answer, we would need to draw a fairly big triangle and be careful with the measuring.

Using more advanced mathematics, it is possible to calculate the value of tan(angle) for any angle. For some, the mathematics you already know is sufficient.

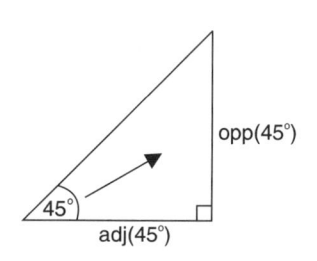

For an angle of 45°, we have to consider a right-angled triangle which has an angle of 45°. The angle sum of any triangle is 180°, so the third angle of our triangle must be $180° - (90° + 45°) = 45°$.

This means that the triangle has two equal angles, and so it is isosceles.

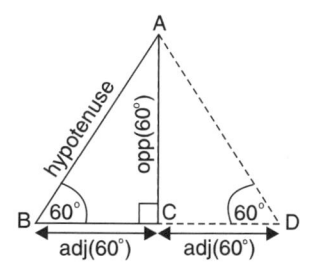

Hence, $\text{opp}(45°) = \text{adj}(45°)$ and so
$\tan(45°) = \frac{\text{opp}(45°)}{\text{adj}(45°)} = 1$

For an angle of 60°, we consider a right-angled triangle ABC with angle B = 60°. Notice that ABC is half an equilateral triangle ABD. Let $\text{adj}(60°) = 1$ unit. Then the side BD of the equilateral triangle is 2 units and so AB = 2 units.

Using Pythagoras's theorem for triangle ABC.
$(AB)^2 = (BC)^2 + (CA)^2$
$2^2 = 1^2 + (CA)^2$
$(CA)^2 = 2^2 - 1^2 = 4 - 1$
$(CA)^2 = 3$
$CA = \sqrt{3}$

We now have $\text{opp}(60°) = \sqrt{3}$ units and $\text{adj}(60°) = 1$ unit, and so
$\tan(60°) = \frac{\text{opp}(60°)}{\text{adj}(60°)} = \frac{\sqrt{3}}{1} = 1.732$ to 4 significant figures

> it is very unlikely that you could obtain tan(60°) as accurately as this from a scale drawing

Using your calculator

You can find the values of tan(angle) in books of mathematical tables. They are also stored in all scientific calculators. We will concentrate on the use of a scientific calculator.

You can obtain the values of tan(angle) on a calculator by using the [tan] key. To find the value of tan 45°, you press the keys [4] [5] [tan].

Notice that you press the [tan] key after you have keyed in the angle.

As you already know, tan(45°) = 1, so your calculator should display

| DEG   1 |

> Make sure that you do get the same answer as me. If you are using a D.A.L. calculator you will need to press the [tan] key *before* you have keyed in the angle and then the [=] key.

If your calculator display is not 1, it is probably because it has found tan(45 *grades*) or tan(45 *radians*) instead of tan(45 *degrees*). The calculator display will be

| GRAD 0.854080685 | or | RAD   1.619775185 |.

(Degrees, grades and radians are different units for measuring angles. 90 degrees = 100 grades = $\frac{1}{2}\pi$ radians.)

For IGCSE mathematics, angles are always measured in degrees. If your calculator is not in 'degrees mode' you must press the appropriate keys so that the display shows 'DEG' before you do any trigonometrical calculations. For the CASIO fx 82 super calculator the keys are [mode] [4].

Example 1

Use you calculator to find:
a) tan(35°)
b) tan(76°)
c) tan(57.3°)
d) tan(89°)

Solution

a) Press the keys [3] [5] [tan]. There is no need to press the [=] key.

The display is | DEG   0.700207538 |

so tan(35°) = 0.700207538.

b) Press the keys [7] [6] [tan].

The display is | DEG   4.010780934 |

so tan(76°) = 4.010780934.

c) Press the keys [5][7][.][3][tan].

The display is | DEG   1.557660082 |

so tan(57.3°) = 1.557660082.

d) Press the keys [8][9][tan].

The display is | DEG   57.28996163 |

so tan(89°) = 57.28996163.

Example 2

Use your calculator to find, correct to 4 decimal places, the value of:
a) tan(5.8°)
b) tan(25°)
c) tan(50°)
d) tan(85°)

Solution

a) Press the keys [5][.][8][tan].

The display is | DEG   0.101576296 |

so tan(5.8°) = 0.1016 to 4 decimal places.

b) Press the keys [2][5][tan].

The display is | DEG   0.466307658 |

so tan(25°) = 0.4663 to 4 decimal places.

c) Press the keys [5][0][tan].

The display is | DEG   1.191753593 |

so tan(50°) = 1.918 to 4 decimal places.

d) Press the keys [8][5][tan].

The display is | DEG   11.4300523 |

so tan(85°) = 11.4301 to 4 decimal places.

Example 3

Use your calculator to find, correct to 4 significant figures, the value of:
a) tan(18.5°)
b) tan(30°)
c) tan(60.4°)
d) tan(71.6°)

Solution

a) Press the keys [1][8][.][5][tan].

The display is | DEG 0.33459532 |

so tan(18.5°) = 0.3346 to 4 significant figures.

b) Press the keys [3][0][tan].

The display is | DEG 0.577350269 |

so tan(30°) = 0.5774 to 4 significant figures.

c) Press the keys [6][0][.][4][tan].

The display is | DEG 1.760318346 |

so tan(60.4°) = 1.760 to 4 significant figures.

d) Press the keys [7][1][.][6][tan].

The display is | DEG 3.006110904 |

so tan(71.6°) = 3.006 to 4 significant figures.

Example 4

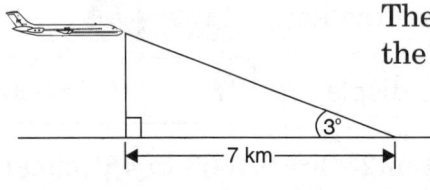

The angle of approach of an aircraft to the runway should be exactly 3°.

How high should the aircraft be when it is 7 km horizontally from the runway?

Give your answer in metres.

Solution

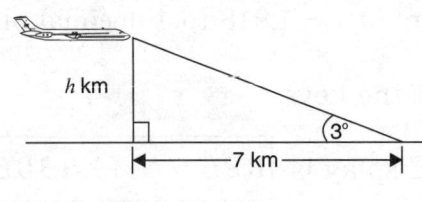

Let the height of the aircraft be h km.

Then $\tan(3°) = \dfrac{\text{opp}(3°)}{\text{adj}(3°)} = \dfrac{h}{7}$

Hence $h = 7 \times \tan(3°)$
$= 7 \times 0.052407779$
$= 0.366854455$

The height of the aircraft
$= 0.366854455$ km
$= 366.854455$ m
$= 367$ m to 3 significant figures

> The sides opp(3°) and adj(3°) must be in the same units. In this example, they are both in kilometres. Because we are told to give the answer in metres, we convert the height from kilometres to metres as the last step.

You should now be ready to obtain and use the tangents of some angles.

EXERCISE 21

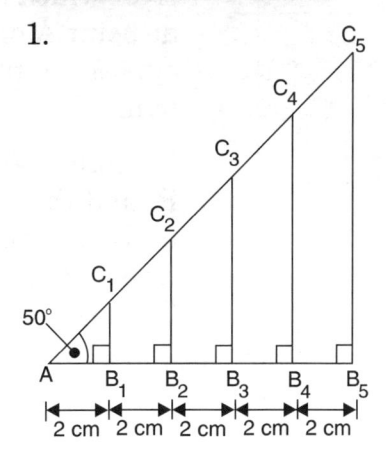

1. This diagram is a sketch of five overlapping right-angled triangles, AB_1C_1, AB_2C_2, AB_3C_3, AB_4C_4, AB_5C_5.

   a) Draw the diagram accurately. (You will find it easier if you use graph paper.)
   b) Measure the lengths of B_1C_1, B_2C_2, B_3C_3, B_4C_4 and B_5C_5.
   c) For each of the five right-angled triangles work out the value of $\frac{\text{opp}(50°)}{\text{adj}(50°)}$ to 2 significant figures.
   d) Use your answer to part c) to estimate the value of $\tan(50°)$.

2. Use your calculator to find, correct to 4 decimal places, the value of:
   a) $\tan(15°)$
   b) $\tan(37.5°)$
   c) $\tan(52.5°)$
   d) $\tan(79°)$

3. Use your calculator to find, correct to 4 significant figures, the value of:
   a) $\tan(1°)$
   b) $\tan(10°)$
   c) $\tan(46.5°)$
   d) $\tan(84.3°)$

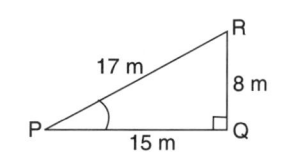

4.
   a) Write down the value of $\tan(P)$ as a fraction.
   b) Find the value of $\tan(P)$ as a decimal correct to 4 places.

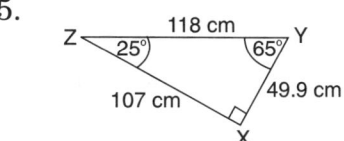

5. From this triangle XYZ, calculate, correct to 3 significant figures, the value of:
   a) $\tan(65°)$
   b) $\tan(25°)$

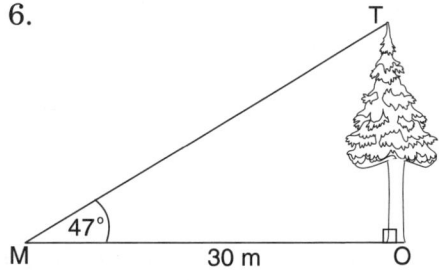

6.
   a) Use your calculator to find the value of $\tan(47°)$ correct to 4 decimal places.
   b) The diagram shows a vertical tree, OT, whose base, O, is 30 m horizontally from the point M. The angle of elevation of T from M is 47°. Calculate the height of the tree.

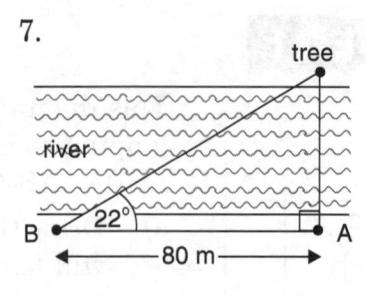

7. Devon wants to estimate the width of a river which has parallel banks. He starts at point A on one of the banks, directly opposite a tree which is on the other bank.

He walks 80 m along the bank to a point B, and then looks back at the tree. He finds that the line between B and the tree makes an angle of 22° with the bank. Calculate the width of the river.

8. tan(45°) = 1. What can you say about the value of tan(A) when A is an acute angle:
   a) less than 45°
   b) greater than 45°

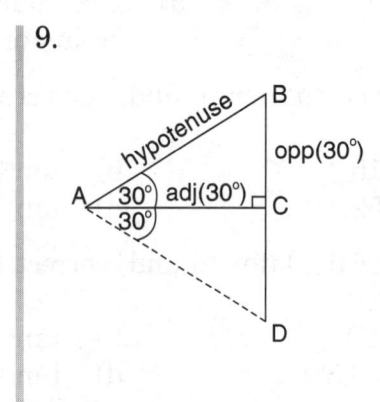

9. The right-angled triangle ABC, in which angle BAC = 30°, can be regarded as half an equilateral triangle ABD.
   a) Taking the length of BC to be 1 unit:
      (i) write down the length of AB
      (ii) use Pythagoras's theorem to obtain the length of AC.
   b) Hence, write down the exact value of $\frac{BC}{AC}$.
   c) Work out the value of tan(30°) correct to 4 significant figures.

Check your answer at the end of this module.

Calculating angles

If you know the size of an angle, you can find the tangent of the angle by using the [tan] key on your calculator. Using the calculator, you can also deal with the 'inverse' (or 'reverse') problem – if you know the tangent of an angle, you can find the size of the angle.

To do this, you need to use the [SHIFT] key followed by the [tan⁻¹] key on the calculator. [tan⁻¹] is an abbreviation for [inverse tan] and it means 'the angle whose tangent is'.

Thus, $\tan^{-1}(2)$ means 'the angle whose tangent is 2'.

To find this angle using the calculator, you press the keys
[2] [SHIFT] [tan⁻¹] .

Provided the calculator is in the 'degrees mode', the display will be

| DEG 63.43494882 | .

This means that the acute angle which has a tangent of 2 is 63.43494882° or, using symbols, $\tan^{-1}(2) = 63.4°$ correct to 1 decimal place.

Example 1

Find, correct to 1 decimal place, the acute angle which has a tangent of:

a) 0.1234
b) 1.0783
c) 5
d) 2.765

Solution

a) Press the keys . 1 2 3 4 SHIFT tan⁻¹ .

The calculator display is DEG 7.034735756

so $\tan^{-1}(0.1234) = 7.0°$ to 1 decimal place.

b) Press the keys 1 . 0 7 8 3 SHIFT tan⁻¹ .

The calculator display is DEG 47.15759935

so $\tan^{-1}(1.0783) = 47.2°$ to 1 decimal place.

c) Press the keys 5 SHIFT tan⁻¹ .

The calculator display is DEG 78.69006753

so $\tan^{-1}(5) = 78.7°$ to 1 decimal place.

d) Press the keys 2 . 7 6 5 SHIFT tan⁻¹ .

The calculator display is DEG 70.11678432

so $\tan^{-1}(2.765) = 70.1°$ to 1 decimal place.

Example 2

Find, correct to 1 decimal place, the acute angle which has a tangent of:

a) $\frac{4}{7}$
b) $\frac{7}{16}$
c) $\frac{27}{19}$
d) $2\frac{10}{13}$

Solution

In each case, the first step is to change the fraction into a decimal. This can be done on the calculator, but you must remember to press the = key to obtain the decimal before you press the SHIFT and tan⁻¹ keys.

a) Press the keys 4 ÷ 7 = SHIFT tan⁻¹ .

The calculator display is DEG 29.7448813

so $\tan^{-1}\left(\frac{4}{7}\right) = 29.7°$ to 1 decimal place.

b) Press the keys [7] [÷] [1] [6] [=] [SHIFT] [tan⁻¹].

The calculator display is $\boxed{DEG\ 23.62937773}$

so $\tan^{-1}\left(\frac{7}{16}\right) = 23.6°$ to 1 decimal place.

c) Press the keys [2] [7] [÷] [1] [9] [=] [SHIFT] [tan⁻¹].

The calculator display is $\boxed{DEG\ 54.86580694}$

so $\tan^{-1}\left(\frac{27}{19}\right) = 54.9°$ to 1 decimal place.

d) You can either change $\frac{10}{13}$ into a decimal and then add 2, or you can say $2\frac{10}{13} = \frac{36}{13}$ and change that into a decimal. We will use the first method:

Press the keys [1] [0] [÷] [1] [3] [=] [+] [2] [=] [SHIFT] [tan⁻¹].

The calculator display is $\boxed{DEG\ 70.14478563}$

so $\tan^{-1}\left(2\frac{10}{13}\right) = 70.1°$ to 1 decimal place.

Example 3

Calculate, correct to 1 decimal place, the lettered angles in these diagrams.

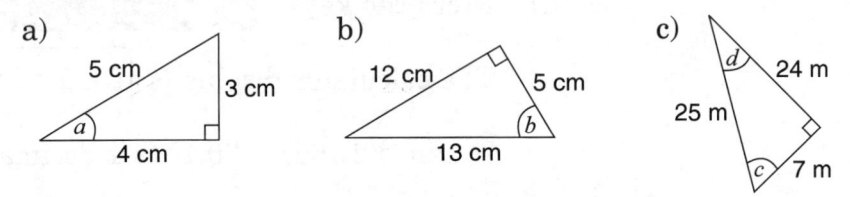

a) [Triangle with sides 5 cm, 3 cm, 4 cm, angle a]

b) [Triangle with sides 12 cm, 5 cm, 13 cm, angle b]

c) [Triangle with sides 25 m, 24 m, 7 m, angles d and c]

Solution

a) $\tan(a) = \frac{\text{opp}(a)}{\text{adj}(a)} = \frac{3}{4} = 0.75$

$a = \tan^{-1}(0.75)$
$= 36.86989765°$ (calculator display)
$a = 36.9°$ to 1 decimal place

b) $\tan(b) = \frac{\text{opp}(b)}{\text{adj}(b)} = \frac{12}{5} = 2.4$

$b = \tan^{-1}(2.4)$
$= 67.38013505°$ (calculator display)
$b = 67.4°$ to 1 decimal place

c) $\tan(c) = \frac{\text{opp}(c)}{\text{adj}(c)} = \frac{24}{7}$

$c = \tan^{-1}\left(\frac{24}{7}\right)$
$= 73.73979529°$ (calculator display)
$c = 73.7°$ to 1 decimal place

To find angle d, we could use the fact that the angle sum of a triangle is 180°. This gives $d = 180° - (90° + 73.7°) = 16.3°$. Alternatively, we could use trigonometry:

$$\tan(d) = \frac{\text{opp}(d)}{\text{adj}(d)} = \frac{7}{24}$$
$$d = \tan^{-1}\left(\frac{7}{24}\right)$$
$$= 16.26020471° \quad \boxed{\text{calculator display}}$$
$$d = 16.3° \text{ to 1 decimal place}$$

Example 4

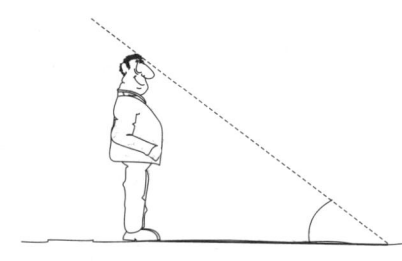

A man, 1.80 m tall, casts a shadow 1.65 m long on horizontal ground.

Calculate the angle of elevation of the sun correct to the nearest degree.

Solution

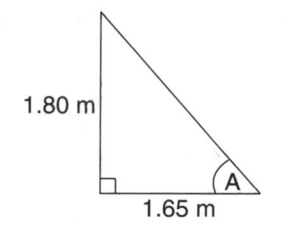

In the diagram, A is the angle of elevation of the sun.

$$\tan(A) = \frac{\text{opp}(A)}{\text{adj}(A)} = \frac{1.80}{1.65}$$
$$A = \tan^{-1}\left(\frac{1.80}{1.65}\right)$$
$$= 47.48955292° \quad \boxed{\text{calculator display}}$$

Angle of elevation of the sun
$= 47°$ to the nearest degree.

I trust that you are beginning to develop an understanding of the tangent ratio. See if you can solve the following problems.

EXERCISE 22

1. Find, correct to 1 decimal place, the acute angle which has a tangent of:
   a) 0.85
   b) 1.2345
   c) 3.56
   d) 10

2. Find, correct to the nearest degree, the acute angle which has a tangent of:
   a) $\frac{2}{5}$
   b) $\frac{7}{9}$
   c) $\frac{25}{32}$
   d) $2\frac{3}{4}$

3. Find, correct to 1 decimal place, the lettered angles in these diagrams.

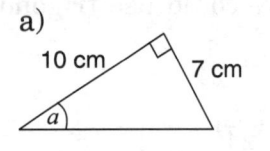

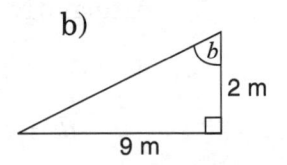

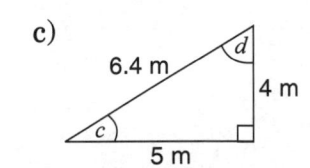

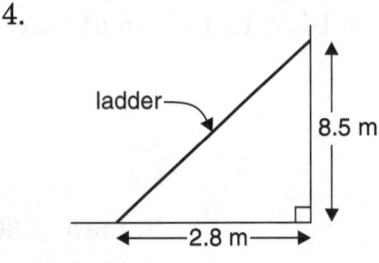

a) 10 cm, 7 cm, angle a

b) b, 2 m, 9 m

c) 6.4 m, d, 4 m, c, 5 m

4.

ladder, 8.5 m, 2.8 m

A ladder stands on horizontal ground and leans against a vertical wall. The foot of the ladder is 2.8 m from the base of the wall and the ladder reaches 8.5 m up the wall.

Calculate the angle that the ladder makes with the ground.

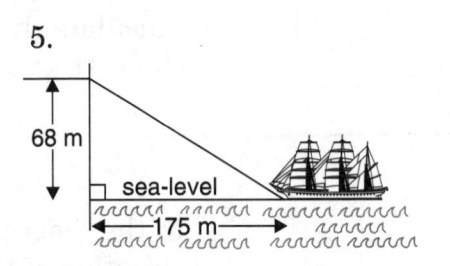

5.

68 m, sea-level, 175 m

The top of a vertical cliff is 68 m above sea-level.

A ship is 175 m from the foot of the cliff.

Calculate the angle of elevation of the top of the cliff from the ship.

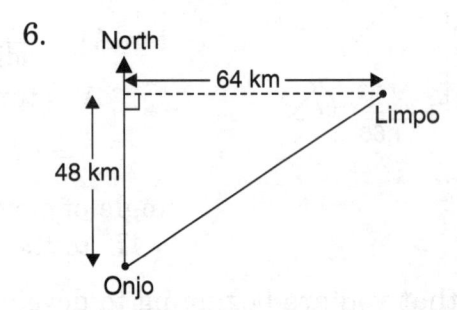

6.

North, 64 km, Limpo, 48 km, Onjo

Limpo is 48 km north and 64 km east of Onjo.

Calculate the 3-figure bearing of Limpo from Onjo.

Have you a good understanding of the tangent ratio? Check your answers at the end of this module.

If you have several wrong answers, you should study the solutions I have given and, if necessary, revise the work on the tangent ratio before you move on to the next section.

C The sine and cosine ratios

As you have seen, the tangent ratio connect two sides of a right-angled triangle. We have called them opp(angle) and adj(angle). The tangent ratio does not involve the hypotenuse of the triangle.

What should we do if the problem we are trying to solve requires us to calculate or to use the hypotenuse?

In one way or another, we could use Pythagoras's theorem:
$[\text{hypotenuse}]^2 = [\text{opp(angle)}]^2 + [\text{adj(angle)}]^2$

But this will not always be easy. It may involve forming and solving an algebraic equation. An alternative is to extend our use of trigonometry. The ratio of two sides of a right-angled triangle is called a trigonometrical ratio, or a **trig ratio**. A trig ratio is independent of the size of the triangle. It depends only on the size of the angles.

There are three trig ratios you need to know for the IGCSE examinations: **tangent**, **sine**, and **cosine**. These names are usually shortened to tan, sin, and cos.

Pronounce 'sine' and 'sin' as 'sign'. Pronounce 'cosine' as 'co-sign' and 'cos' as 'coz'.

You already know that:
$$\tan(\text{angle}) = \frac{\text{opp(angle)}}{\text{adj(angle)}}$$

The definitions of the other two are:
$$\sin(\text{angle}) = \frac{\text{opp(angle)}}{\text{hypotenuse}} \text{ and } \cos(\text{angle}) = \frac{\text{adj(angle)}}{\text{hypotenuse}}$$

For acute angles, the value of tan(angle) can be any size, but the values of sin(angle) and cos(angle) must be less than 1. (This is because the hypotenuse is the longest side in the triangle.)

We use the sine and cosine ratios in a similar way to the tangent ratio. For acute angles, their values can be obtained from a scientific calculator by using the [sin] and [cos] keys respectively.

To find the size of an angle when its sine is known, we use the [SHIFT] and [sin⁻¹] keys on the calculator.

Similarly, to find the size of an angle when its cosine is known, we use the [SHIFT] and [cos⁻¹] keys on the calculator.

Example 1

For each of these triangles, write down the value of:
(i) sin(A)   (ii) cos(A)

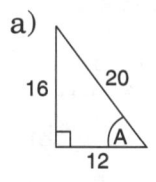

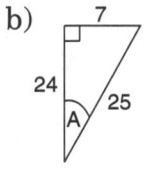

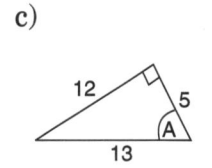

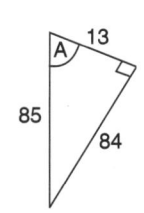

a)   [triangle: sides 16, 20, 12, angle A]
b)   [triangle: sides 7, 24, 25, angle A]
c)   [triangle: sides 12, 13, 5, angle A]
d)   [triangle: sides 13, 85, 84, angle A]

Solution

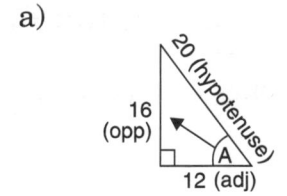

a) [triangle: 16 (opp), 20 (hypotenuse), 12 (adj), angle A]

(i) $\sin(A) = \dfrac{\text{opp}(A)}{\text{hypotenuse}} = \dfrac{16}{20}$

(ii) $\cos(A) = \dfrac{\text{adj}(A)}{\text{hypotenuse}} = \dfrac{12}{20}$

b)

(i) $\sin(A) = \dfrac{\text{opp}(A)}{\text{hypotenuse}} = \dfrac{7}{25}$

(ii) $\cos(A) = \dfrac{\text{adj}(A)}{\text{hypotenuse}} = \dfrac{24}{25}$

c)

(i) $\sin(A) = \dfrac{\text{opp}(A)}{\text{hypotenuse}} = \dfrac{12}{13}$

(ii) $\cos(A) = \dfrac{\text{adj}(A)}{\text{hypotenuse}} = \dfrac{5}{13}$

d)

(i) $\sin(A) = \dfrac{\text{opp}(A)}{\text{hypotenuse}} = \dfrac{84}{85}$

(ii) $\cos(A) = \dfrac{\text{adj}(A)}{\text{hypotenuse}} = \dfrac{13}{85}$

Example 2

Use your calculator to find, correct to 4 decimal places, the value of:
a) $\sin(10°)$
b) $\sin(45°)$
c) $\sin(69.5°)$
d) $\sin(80°)$

Solution

Make sure that your calculator is in the 'degrees mode'.

a) Press the keys [1] [0] [sin].

The display is | DEG   0.173648178 |

so $\sin(10°) = 0.1736$ to 4 decimal places.

b) Press the keys [4] [5] [sin].

The display is | DEG   0.707106781 |

so $\sin(45°) = 0.7071$ to 4 decimal places.

c) Press the keys [6] [9] [.] [5] [sin].

The display is | DEG   0.936672189 |

so $\sin(69.5°) = 0.9367$ to 4 decimal places.

d) Press the keys [8] [0] [sin].

The display is | DEG   0.984807753 |

so $\sin(80°) = 0.9848$ to 4 decimal places.

Example 3

Use your calculator to find, correct to 4 decimal places, the value of:
a) $\cos(10°)$
b) $\cos(45°)$
c) $\cos(20.5°)$
d) $\cos(80°)$

Solution

a) Press the keys [1] [0] [cos].

The display is | DEG   0.984807753 |

so cos(10°) = 0.9848 to 4 decimal places.

b) Press the keys [4] [5] [cos].

The display is | DEG   0.707106781 |

so sin(45°) = 0.7071 to 4 decimal places.

c) Press the keys [2] [0] [.] [5] [cos].

The display is | DEG   0.936672189 |

so cos(20.5°) = 0.9367 to 4 decimal places.

d) Press the keys [8] [0] [cos].

The display is | DEG   0.173648178 |

so cos(80°) = 0.1736 to 4 decimal places.

Did you notice that sin(10°) = cos(80°), sin(45°) = cos(45°), sin(69.5°) = cos(20.5°) and sin(80°) = cos(10°)?

Can you see a reason for this?

What is the connection between the angles in each pair?

Example 4

For each of these triangles, write down the ratio of sides corresponding to the trig ratio named.

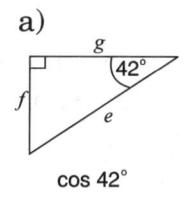

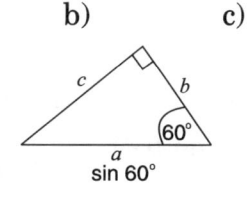

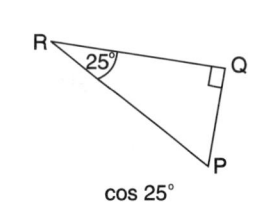

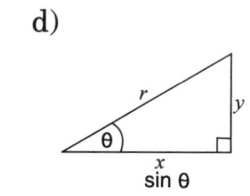

a) cos 42° b) sin 60° c) cos 25° d) sin θ

Solution

a) $\cos(42°) = \dfrac{\text{adj}(42°)}{\text{hypotenuse}} = \dfrac{g}{e}$

b) $\sin(60°) = \dfrac{\text{opp}(60°)}{\text{hypotenuse}} = \dfrac{c}{a}$

c) $\cos(25°) = \dfrac{\text{adj}(25°)}{\text{hypotenuse}} = \dfrac{QR}{PR}$

d) $\sin(\theta) = \dfrac{\text{opp}(\theta)}{\text{hypotenuse}} = \dfrac{y}{r}$

Example 5

For each of these triangles, calculate, correct to 3 significant figures, the length of the lettered side.

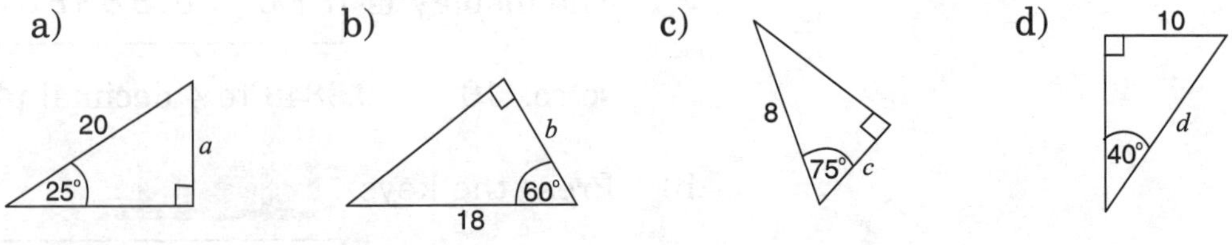

a)   b)   c)   d)

Solution

a) The lettered side is *opposite* the angle of 25°, so we use

$$\sin(25°) = \frac{\text{opp}(25°)}{\text{hypotenuse}}$$

$$\sin(25°) = \frac{a}{20}$$

$$a = 20 \times \sin(25°)$$
$$= 8.452365235$$

$\boxed{2}\boxed{0}\boxed{\times}\boxed{2}\boxed{5}\boxed{\sin}\boxed{=}$

$a = 8.45$ to 3 significant figures

b) The lettered side is *adjacent* to the angle of 60°, so we use

$$\cos(60°) = \frac{\text{adj}(60°)}{\text{hypotenuse}}$$

$$\cos(60°) = \frac{b}{18}$$

$$b = 18 \times \cos(60°)$$

$\boxed{1}\boxed{8}\boxed{\times}\boxed{6}\boxed{0}\boxed{\cos}\boxed{=}$

$$b = 9$$

c) The lettered side is *adjacent* to the angle of 75°, so we use

$$\cos(75°) = \frac{\text{opp}(75°)}{\text{hypotenuse}}$$

$$\cos(75°) = \frac{c}{8}$$

$$c = 8 \times \cos(75°)$$
$$= 2.070552361$$

$\boxed{8}\boxed{\times}\boxed{7}\boxed{5}\boxed{\cos}\boxed{=}$

$c = 2.07$ to 3 significant figures

d) The lettered side is the *hypotenuse* and the known side (10) is *opposite* the angle of 40°, so we use

$$\sin(40°) = \frac{\text{opp}(40°)}{\text{hypotenuse}}$$

$$\sin(40°) = \frac{10}{d}$$

$$d \times \sin(40°) = 10$$

$$d = \frac{10}{\sin(40°)}$$
$$= 15.55723827$$

$\boxed{1}\boxed{0}\boxed{\div}\boxed{4}\boxed{0}\boxed{\sin}\boxed{=}$

$d = 15.6$ to 3 significant figures

The work in part d) is more difficult than in the other examples because the unknown length appears as the denominator of the fraction. This means that some algebraic manipulation is required to find the unknown length.

Learners often make mistakes here and obtain answers which are too small. You should always check that your answer is sensible. In particular check that the hypotenuse is the longest side in the right-angled triangle.

Example 6

Use your calculator to find, correct to 1 decimal place:
a) the acute angle whose sine is 0.35
b) the acute angle whose cosine is 0.5461
c) the acute angle whose sine is $\frac{5}{8}$
d) the acute angle whose cosine is $\frac{12}{19}$

Solution

a) Press the keys [.] [3] [5] [SHIFT] [sin⁻¹] .

   The display is | DEG   20.48731512 |

   so $\sin^{-1}(0.35) = 20.5°$ to 1 decimal place.

b) Press the keys [.] [5] [4] [6] [1] [SHIFT] [cos⁻¹] .

   The display is | DEG   65.90013412 |

   so $\cos^{-1}(0.5461) = 56.9°$ to 1 decimal place.

c) Press the keys [5] [÷] [8] [=] [SHIFT] [sin⁻¹] .

   The display is | DEG   38.68218745 |

   so $\sin^{-1}(\frac{5}{8}) = 38.7°$ to 1 decimal place.

d) Press the keys [1] [2] [÷] [1] [9] [=] [SHIFT] [cos⁻¹] .

   The display is | DEG   50.83328928 |

   so $\cos^{-1}(\frac{12}{19}) = 50.8°$ to 1 decimal place.

Example 7

Find, correct to the nearest degree, the size of the lettered angle in each of the following diagrams.

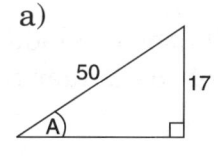

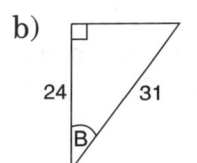

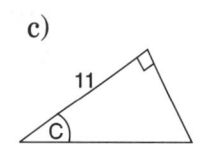

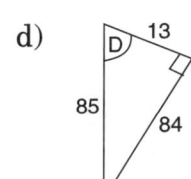

a) [triangle with sides 50, 17, angle A]
b) [triangle with sides 24, 31, angle B]
c) [triangle with sides 11, 13, angle C]
d) [triangle with sides 13, 85, 84, angle D]

Solution

a) We know that hypotenuse = 50 and opp(A) = 17, so we use
$$\sin(A) = \frac{\text{opp}(A)}{\text{hypotenuse}} = \frac{17}{50} = 0.34$$
$$A = \sin^{-1}(0.34)$$
$$= 19.877687407° \quad \boxed{\text{calculator display}}$$
$$A = 20° \text{ to the nearest degree}$$

b) We know that hypotenuse = 31 and adj(B) = 24, so we use
$$\cos(B) = \frac{\text{adj}(B)}{\text{hypotenuse}} = \frac{24}{31}$$
$$B = \cos^{-1}\left(\frac{24}{31}\right) \quad \boxed{2}\boxed{4}\boxed{\div}\boxed{3}\boxed{1}\boxed{=}\boxed{\text{SHIFT}}\boxed{\cos^{-1}}$$
$$= 39.268026°$$
$$B = 39° \text{ to the nearest degree}$$

c) We use $\cos(C) = \dfrac{\text{adj}(C)}{\text{hypotenuse}} = \dfrac{11}{13}$
$$C = \cos^{-1}\left(\frac{11}{13}\right) \quad \boxed{1}\boxed{1}\boxed{\div}\boxed{1}\boxed{3}\boxed{=}\boxed{\text{SHIFT}}\boxed{\cos^{-1}}$$
$$= 32.20422751°$$
$$C = 32° \text{ to the nearest degree}$$

d) We have a choice here.
We could use
$$\sin(D) = \frac{\text{opp}(D)}{\text{hypotenuse}} = \frac{84}{85}$$
$$D = \sin^{-1}\left(\frac{84}{85}\right)$$
$$= 81.20258929°$$
$$D = 81° \text{ to the nearest degree}$$

Alternatively, we could use $\cos(D) = \dfrac{\text{adj}(D)}{\text{hypotenuse}} = \dfrac{13}{85}$

or $\tan(D) = \dfrac{\text{opp}(D)}{\text{adj}(D)} = \dfrac{84}{85}$

Use your calculator to check that these methods also give $D = 81.20258929°$.

Example 8

A ladder, 4.8 m long, leans against a vertical wall with its foot on horizontal ground. The ladder makes an angle of 70° with the ground.
a) How far up the wall does the ladder reach?
b) How far is the foot of the ladder from the wall?

Solution

In the diagram, AC is the hypotenuse of the right-angled triangle ABC, AB = opp(70°) and BC = adj(70°)

a) $\sin(70°) = \dfrac{\text{opp}(70°)}{\text{hypotenuse}} = \dfrac{AB}{4.8}$

$AB = 4.8 \times \sin(70°)$
$= 4.51052458$

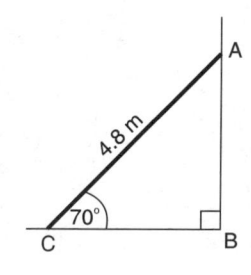

The ladder reaches 4.51 m up the wall.

b) $\cos(70°) = \dfrac{\text{adj}(70°)}{\text{hypotenuse}} = \dfrac{BC}{4.8}$

$BC = 4.8 \times \cos(70°)$
$= 1.641696688$

The foot of the ladder is 1.64 m from the wall (to 3 sig. figures).

> We could have used Pythagoras's theorem to calculate BC from the lengths of AC and AB. However, it is better to use trigonometry (in case we had made a mistake in calculating AB).

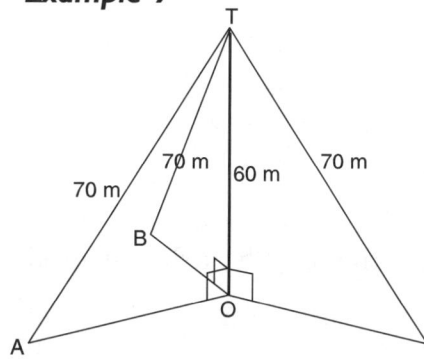

Example 9

A television transmission mast, OT, is vertical and stands on horizontal ground.

It is 60 m high and is supported by three wires, AT, BT and CT, each 70 m long, as shown in the diagram.

Find the angle that the supporting wires make with the horizontal.

Solution

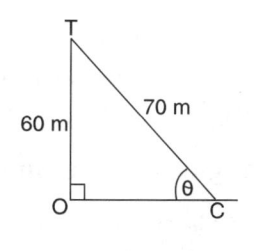

This diagram shows the mast, OT, and one of the supporting wires, CT.

The mast is vertical and the ground is horizontal, so angle TOC = 90°.

The angle CT makes with the horizontal = θ.

$\sin(\theta) = \dfrac{\text{opp}(\theta)}{\text{hypotenuse}} = \dfrac{60}{70}$

$\theta = \sin^{-1}\left(\dfrac{60}{70}\right)$

$= 58.99728087$ [calculator display]

The angle each wire makes with the horizontal = 59.0° to 1 decimal place.

We have covered quite a lot of ground since the last exercise and it is now time to see whether you have grasped all the ideas.

EXERCISE 23

1. For each of these triangles, write down the value of:
   (i) sin(A)   (ii) cos(A)

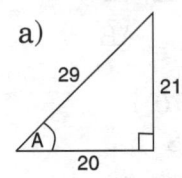

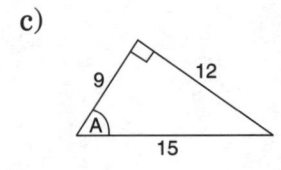

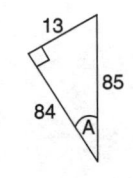

   a) [triangle with sides 29, 21, 20, right angle, angle A at bottom left]
   b) [triangle with sides 8, 17, 15, right angle, angle A]
   c) [triangle with sides 9, 12, 15, angle A at bottom left]
   d) [triangle with sides 13, 85, 84, right angle, angle A]

2. Use your calculator to find, correct to 4 decimal places, the value of:
   a) sin(5°)   b) sin(30°)
   c) sin(60°) d) sin(85°)

3. Use your calculator to find, correct to 4 decimal places, the value of:
   a) cos(5°)   b) cos(30°)
   c) cos(60°) d) cos(85°)

4. For each of these triangles, write down the ratio of the sides corresponding to the trig ratio named.

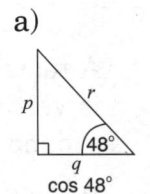

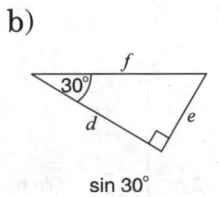

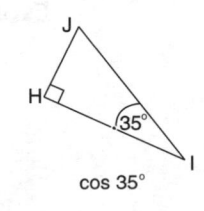

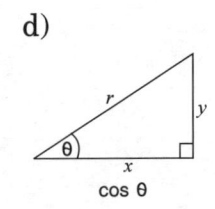

   a) [triangle with sides p, q, r, angle 48°] cos 48°
   b) [triangle with sides d, e, f, angle 30°] sin 30°
   c) [triangle HJI with angle 35° at J] cos 35°
   d) [triangle with sides r, x, y, angle θ] cos θ

5. For each of these triangles, calculate, correct to 3 significant figures, the length of the lettered side.

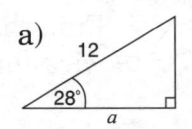

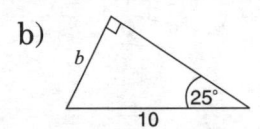

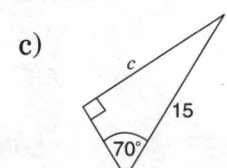

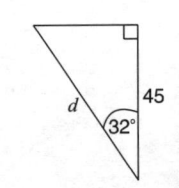

   a) [triangle with side 12, angle 28°, side a]
   b) [triangle with side b, angle 25°, side 10]
   c) [triangle with side c, 15, angle 70°]
   d) [triangle with side d, 45, angle 32°]

6. Use your calculator to find, correct to 1 decimal place:
   a) the acute angle whose sine is 0.99
   b) the acute angle whose cosine is 0.5432
   c) the acute angle whose sine is $\frac{3}{8}$
   d) the acute angle whose cosine is $\frac{10}{23}$

7. Find, correct to the nearest degree, the size of the lettered angle in each of the following diagrams.

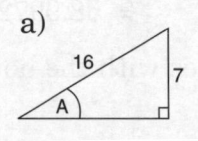

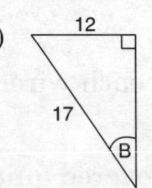

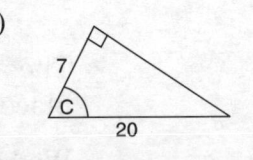

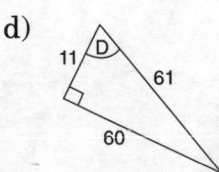

   a) [triangle with sides 16, 7, angle A]
   b) [triangle with sides 12, 17, angle B]
   c) [triangle with sides 7, 20, angle C]
   d) [triangle with sides 11, 61, 60, angle D]

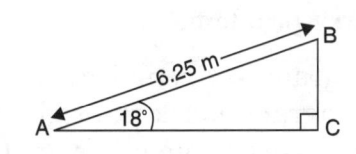

8. The diagram shows a ramp, AB, which makes an angle of 18° with the horizontal. The ramp is 6.25 m long.

Calculate the difference in height between A and B. (This is the length of BC in the diagram.)

9. Village Q is 18 km from village P, on a bearing of 056°.
   a) Calculate the distance Q is north of P.
   b) Calculate the distance Q is east of P.

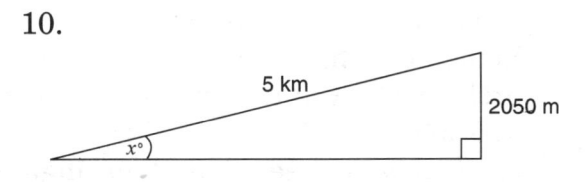

10. A mountain railway track is 5 km long. The top of the track is 2050 m above the lower end.

Calculate x, the angle of slope of the track.

It may have taken you quite a long time to answer all the questions in Exercise 23. Do you feel that you did well? Check your answers at the end of this module.

D Using trigonometry to solve problems

When you are told to *calculate* lengths and/or angles in a diagram, you must *not* use scale drawing. Solving such problems often involves using Pythagoras's theorem and/or trigonometry.

You may have to use one or more of the trig ratios. You will have to decide whether to use the sine, cosine and/or tangent. It is essential that you remember the correct definitions:

$$\text{sine} = \frac{\text{opposite}}{\text{hypotenuse}}, \quad \text{cosine} = \frac{\text{adjacent}}{\text{hypotenuse}}, \quad \text{tangent} = \frac{\text{opposite}}{\text{adjacent}}$$

Some people use the 'words' SOH, CAH, TOA to remind them of the definitions. SOH stands for $S = \frac{O}{H}$ which is $\text{sine} = \frac{\text{opposite}}{\text{hypotenuse}}$.

Can you work out the cosine and tangent ratios using CAH and TOA?

A silly sentence like this: '**S**am **O**ften **H**as **C**offee **A**nd **H**amburgers **T**o **O**vercome **A**nger' can help you to remember SOH CAH TOA.

You can make up your own silly sentence but, somehow or another, you must remember the definitions correctly. If you use an incorrect definition when you are answering an examination question, you are wasting your time. You will get no marks!

Here is some advice to help you when you are tackling a trigonometry problem:

- If no diagram is given, draw one yourself.
- In the diagram, mark the right angles.
- Show the sizes of the other angles which are known and the lengths of any lines which are known.
- Mark the angles or sides which you have to calculate.
- Identify the right-angled triangle(s) which contain(s) the angles or sides you have to calculate.
- If it is a 3-dimensional problem, draw separate diagrams of the right-angled triangles you are going to use.
- Consider whether you need to create right-angled triangles by drawing extra lines in your diagram(s). For example, an isosceles triangle can be divided into two congruent right-angled triangles.
- Decide on the steps you will take to solve the problem – will you need Pythagoras's theorem, sine, cosine and/or tangent?
- Remember that there is often more than one way of solving a problem – if you can, choose the shortest method using the given information.
- If you have to use a length or angle which you have already calculated, use the most accurate value which you have – do not use a 'rounded' value.
- Show how you obtained your answer(s) – in examinations, marks are given for correct methods as well as for correct answers.
- Give sensible answer(s) to the problem – usually 3-figure accuracy for lengths and 1 decimal place accuracy for angles (unless the answer is exact). Remember that your calculator will give many more figures than is sensible in a practical problem.
- Check that your answer is reasonable – remember that the hypotenuse is the longest side in a right-angled triangle, and that the shortest side is opposite the smallest angle.

Example 1

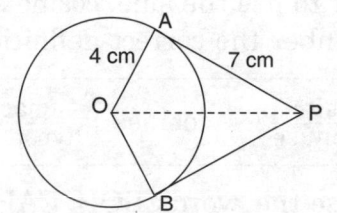

Tangents are drawn from P to touch the circle, centre O, at A and B. PA = 7 cm and OA = 4 cm.

Calculate angle APB.

Solution

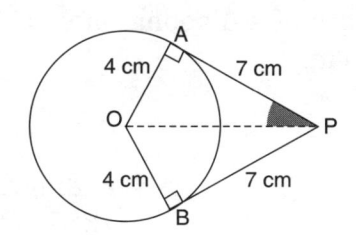

The angle between a tangent and the radius to the point of contact is 90°, so angle OAP = 90° and angle OBP = 90°.

Join O to P to form two congruent, right-angled triangles OAP and OBP.

Let angle APO = x and angle APB = y

In triangle APO, $\tan x = \frac{\text{opp}(x)}{\text{adj}(x)} = \frac{4}{7}$

$$x = \tan^{-1}\left(\frac{4}{7}\right)$$
$$= 29.7448813°$$

$$y = 2 \times x$$
$$= 59.4897626°$$
$$y = 59.5° \text{ to 1 decimal place}$$

Example 2

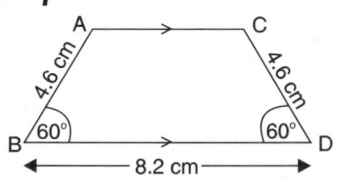

The diagram represents an isosceles trapezium ABCD.

Calculate the area of the trapezium.

Solution

The area of a trapezium = (average of the parallel sides) × (perpendicular distance between them)

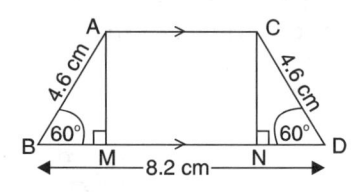

In this diagram, we need to calculate the lengths of AM (or CN) and AC.

AC = MN and we can find the length of MN if we calculate the lengths of BM and ND.

In triangle ABM,

$\sin(60°) = \frac{\text{opp}(60°)}{\text{hypotenuse}} = \frac{\text{AM}}{4.6}$ and $\cos(60°) = \frac{\text{adj}(60°)}{\text{hypotenuse}} = \frac{\text{BM}}{4.6}$

Hence, AM = 4.6 × sin(60°) and BM = 4.6 × cos(60°)
   AM = 3.983716857 cm and BM = 2.3 cm

By symmetry, ND = BM = 2.3 cm and
so MN = 8.2 − (2.3 + 2.3) = 3.6 cm

Hence, AC = 3.6 cm and AM = CN = 3.983716857 cm

The area of ABCD = $\left(\frac{\text{AC} + \text{BD}}{2}\right) \times \text{AM}$

$$= \left(\frac{3.6 + 8.2}{2}\right) \times 3.983716857 \text{ cm}^2$$
$$= 23.5039296 \text{ cm}^2$$

Area of ABCD = 23.5 cm² to 3 significant figures.

Example 3

Find the length of a diagonal of a regular pentagon which has sides of length 10 cm.

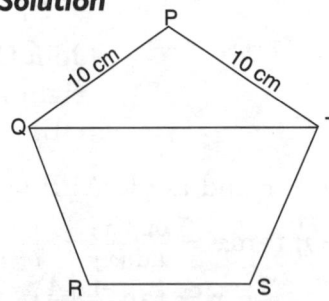

Solution

The diagram shows a regular pentagon PQRST. QT is one of the diagonals.

Angle sum of a pentagon
$= (5 \times 180°) - 360°$
$= 540°$

Angle QPT $= \frac{540°}{5} = 108°$

We create congruent right-angled triangles by drawing the perpendicular PN from P to QT.

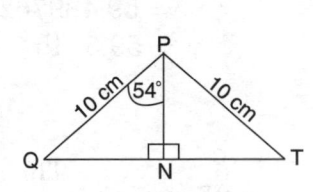

By symmetry,
angle QPN = angle TPN $= \frac{108°}{2} = 54°$
and QN = NT

In triangle PQN,
$\sin(54°) = \frac{\text{opp}(54°)}{\text{hypotenuse}} = \frac{QN}{10}$

$QN = 10 \times \sin(54°)$
$= 8.090169944$

Hence, QT $= 2 \times QN = 16.18033989$

The length of each diagonal = 16.2 cm to 3 significant figures.

Example 4

Calculate the angles of an isosceles triangle which has sides of length 9 cm, 9 cm and 14 cm.

Solution

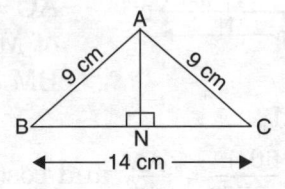

The diagram shows the isosceles triangle ABC.

AN is the perpendicular from A to BC and it is the line of symmetry of the triangle.
BN = NC = 7 cm.

In triangle ABN, $\cos(B) = \frac{\text{adj}(B)}{\text{hypotenuse}} = \frac{BN}{AB} = \frac{7}{9}$

$B = \cos^{-1}\left(\frac{7}{9}\right)$

$B = 38.94244127°$
Similarly $C = 38.94244127°$
Hence angle BAC $= 180° - (B + C) = 102.1151175°$

Correct to 1 decimal place, the angles of the isosceles triangle ABC are 102.1°, 38.9° and 38.9°.

These values do not add up to 180° because each of them is approximate – that is, correct to 1 decimal place. If we give them correct to the nearest degree, they are 102°, 39° and 39°, which do add up to 180°!

Example 5

A plane starts from P and flies 50 km on a bearing of 060° to Q, then 75 km on a bearing 160° to R.
a) How far east of P is R?
b) How far south of P is R?
c) What is the bearing of P from R?

Solution

Figure 1 shows the information given in the question.

To solve the problem, we create right-angled triangles PQN and QRM by drawing east-west lines through Q and R, as shown in Figure 2.

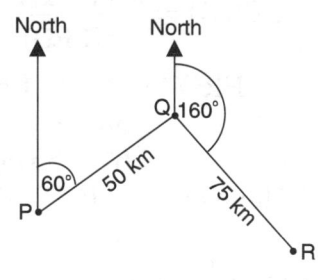

Figure 1

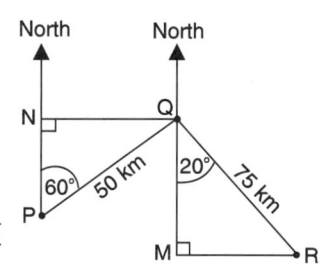

Figure 2

Angle RQM = 180° − 160°
= 20°

a) The distance R is east of P = NQ + MR
In triangle PQN, $\sin(60°) = \frac{\text{opp}(60°)}{\text{hypotenuse}} = \frac{NQ}{50}$
so NQ = 50 × sin(60°)
= 43.301 . . . km

In triangle RQM, $\sin(20°) = \frac{\text{opp}(20°)}{\text{hypotenuse}} = \frac{MR}{75}$
so MR = 75 × sin(20°)
= 25.651 . . . km

Distance R is east of P = (43.301 . . . + 25.651 . . .) km
= 68.952 . . . km

Distance R is east of P = 69.0 km to 3 significant figures.

b) The distance R is south of P = QM − NP
In triangle PQN, $\cos(60°) = \frac{\text{adj}(60°)}{\text{hypotenuse}} = \frac{NP}{50}$
so NP = 50 × cos(60°)
= 25 km

In triangle RQM, $\cos(20°) = \frac{\text{adj}(20°)}{\text{hypotenuse}} = \frac{QM}{75}$
so QM = 75 × cos(20°)
= 70.4769 . . . km

Distance R is south of P = (70.4769 . . . − 25) km
= 45.4769 . . . km

Distance R is south of P = 45.5 km to 3 significant figures.

c)

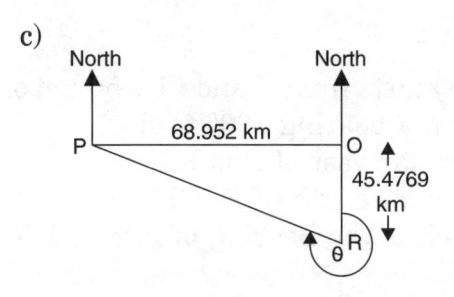

The bearing of P *from R* is measured *at R* in a clockwise direction from the north. It is the angle θ in this diagram.

The line PO is east-west and the line RO is north-south.
$\theta = 360° - $ angle PRO.

In triangle PRO, $\tan(\hat{PRO}) = \dfrac{\text{opp}(\hat{PRO})}{\text{adj}(\hat{PRO})} = \dfrac{68.952}{45.4769} = 1.5162$

$$\hat{PRO} = \tan^{-1}(1.5162)$$
$$= 56.59°$$
$$= 57° \text{ to the nearest degree}$$
$$\text{Hence } \theta = 360° - 57°$$
$$= 303°$$

The bearing of P from R = 303°.

Example 6

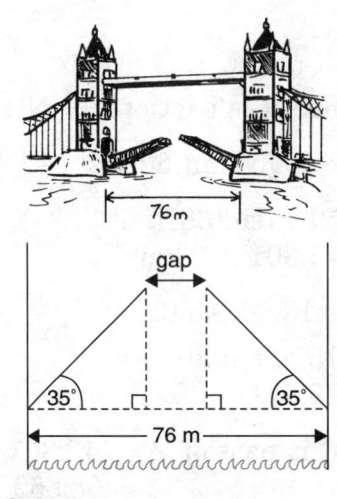

This is Tower Bridge in London.

The span between the tower is 76 m.

This is a simplified drawing of the bridge showing the two halves raised to 35°.

How wide is the gap?

Solution

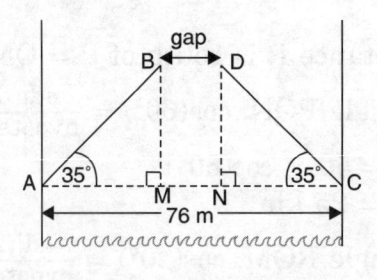

The gap = BD = MN and MN = AC − (AM + NC).

The right-angled triangles ABM and CDN are congruent, so AM = NC.

When the two halves are lowered, they must meet in the middle,

so AB = CD = $\dfrac{76 \text{ m}}{2}$ = 38 m.

In the triangle ABM, $\cos(35°) = \dfrac{\text{adj}(35°)}{\text{hypotenuse}} = \dfrac{\text{AM}}{38}$

$$\text{AM} = 38 \times \cos(35°)$$
$$= 31.1277\ldots \text{ m}$$
$$\text{Hence MN} = (76 - (31.1277\ldots + 31.1277\ldots)) \text{ m}$$
$$= 13.744\ldots \text{ m}$$

The gap BD = 13.7 m to 3 significant figures.

Here are some trigonometry questions for you to try.
Before you start, look again at the advice given on page 116.

EXERCISE 24

1. The diagram represents a ramp AB for a lifeboat. AC is vertical and CB is horizontal.

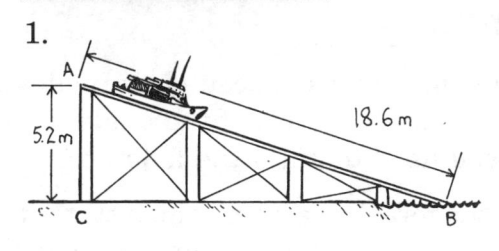

   a) Calculate the size of the angle ABC correct to 1 decimal place.

b) Calculate the length of BC correct to 3 significant figures.

2. AB is a chord of a circle, centre O, radius 8 cm.
   Angle AOB = 120°.

   Calculate the length of AB.

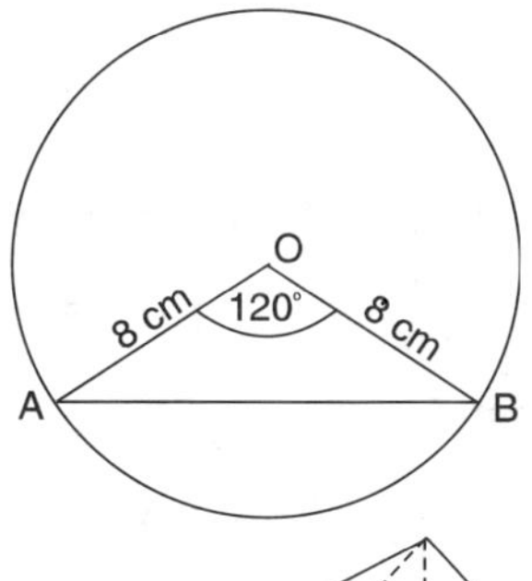

3. The diagram represents a tent which is in the shape of a triangular prism. The front of the tent, ABD, is an isosceles triangle with AB = AD.

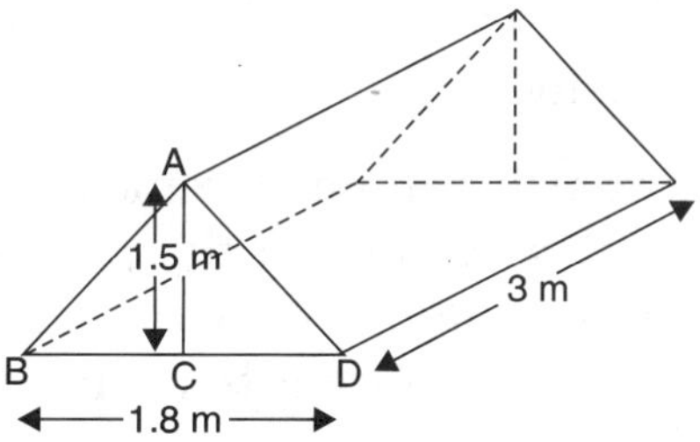

   The width, BD, is 1.8 m and the supporting pole AC is perpendicular to BD and 1.5 m high. The tent is 3 m long.

   Calculate:
   a) the angle between AB and BD
   b) the length of AB
   c) the volume inside the tent

4. The sketch represents a field PQRS on level ground.
   The sides PQ and SR run due east.
   a) Write down the bearing of S from P.
   b) Calculate the shortest distance between SR and PQ.
   c) Calculate, in square metres, the area of the field PQRS.

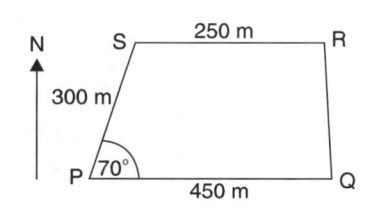

5. In the isosceles triangle DEF, angle E = angle F = 35° and side EF = 10 cm.
   a) Calculate the perpendicular distance from D to EF.
   b) Calculate the length of the side DE.

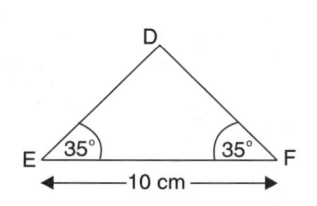

By now you should have some idea of the variety of problems which can be solved by trigonometry. Were you able to deal with all the questions in Exercise 24? Check your answers at the end of this module.

Summary

In this unit we have been concerned with calculating the length of the sides and the size of the angles in right-angled triangles. To do this you need to know the following:

- Pythagoras's theorem, which states that in a right-angled triangle, the square on the hypotenuse is equal to the sum of the squares on the other two sides. So in this triangle, $a^2 = b^2 + c^2$

- $\tan(\text{angle}) = \frac{\text{opp(angle)}}{\text{adj(angle)}}$ and if you know what the tangent of the angle is you can use $\tan^{-1}$ to find the size of the angle

- $\sin(\text{angle}) = \frac{\text{opp(angle)}}{\text{hypotenuse}}$ and $\sin^{-1}$ will give you the angle if you know its sine

- $\cos(\text{angle}) = \frac{\text{adj(angle)}}{\text{hypotenuse}}$ and $\cos^{-1}$ will give you the angle if you know its cosine.

You can see that if you don't remember the trig ratios properly you won't be able to solve any of the problems correctly. I showed you how to remember the ratios by remembering SOH CAH TOA. Learn the trig ratios well before attempting the 'Check your progress.'

Check your progress

1. Evaluate $15\cos(40°)$
   a) Write down the answer as accurately as your calculator will allow.
   b) Round off your answer to 3 significant figures.

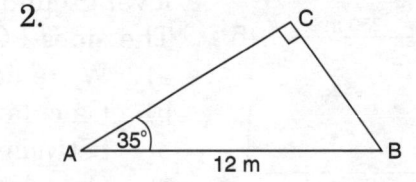

2. The diagram shows the cross-section of the roof of Mr. Haziz's house. The house is 12 m wide, angle CAB = 35° and angle ACB = 90°.

   Calculate the lengths of the two sides of the roof, AC and BC.

3. The diagram shows a trapezium ABCD in which angle ABC = angle BCD = 90°, AB = 90 mm, BC = 72 mm and CD = 25 mm.

   Calculate the perimeter of the trapezium.

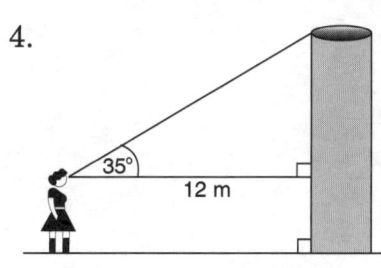

4. A girl, whose eyes are 1.5 m above the ground, stands 12 m away from a tall chimney. She has to raise her eyes 35° upwards from the horizontal to look directly at the top of the chimney.

Calculate the height of the chimney.

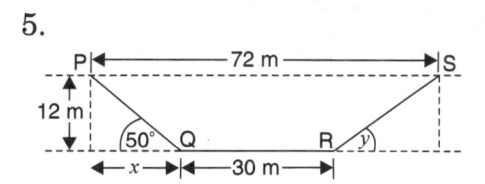

5. The diagram shows the cross-section PQRS of a cutting made for a road. PS and QR are horizontal. PQ makes an angle of 50° with the horizontal.

   a) Calculate the horizontal distance between P and Q (marked x in the diagram).
   b) Calculate the angle which RS makes with the horizontal (marked y in the diagram).

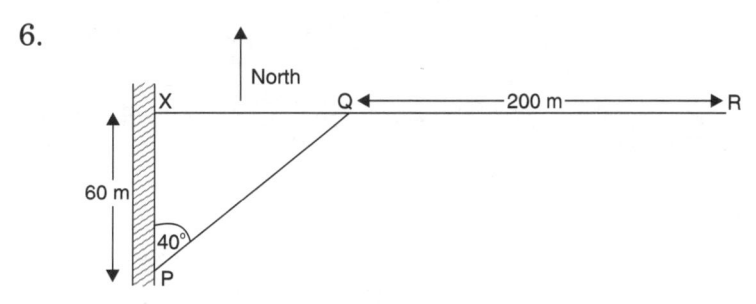

6. A game warden is standing at a point P alongside a road which runs north-south. There is a marker post at the point X, 60 m north of his position. The game warden sees a lion at Q on a bearing of 040° from him and due east of the marker post.

   a) (i) Show by calculation that the distance, QX, of the lion from the road is 50.3 m correct to 3 significant figures.
      (ii) Calculate the distance, PQ, of the lion from the game warden.
   b) Another lion appears at R, 200 m due east of the first one at Q.
      (i) Write down the distance XR.
      (ii) Calculate the distance, PR, of the second lion from the game warden.
      (iii) Calculate the bearing of the second lion from the game warden, correct to the nearest degree.

Check your answers at the end of this module.

You have now reached the end of Unit 3.

If you are following the IGCSE CORE syllabus, you have completed all the work you need to do in Module 5. I hope you feel that you have a good grasp of the work on perimeters, areas, volumes, Pythagoras's theorem and trigonometry.

If you are intending to take the EXTENDED syllabus examinations, you need to study trigonometry a little further. You will find the extra topics covered in Unit 4.

Unit 4
Trigonometry Extended

In this unit you will extend your knowledge of trigonometry so that you are not limited to using trigonometry only in right-angled triangles. You'll also learn about the sine and cosine functions.

This unit is divided into three sections:

Section	Title	Time
A	Trigonometry for any triangle	4 hours
B	The sine and cosine functions	2 hours
C	Further applications of trigonometry	3 hours

By the end of this unit, you should be able to:

- find the sine and cosine of obtuse angles
- use the formula $\frac{1}{2}ab\sin C$ for the area of a triangle
- use the sine rule and the cosine rule to calculate sides and angles of triangles
- draw graphs of sine and cosine functions
- calculate the angle between a line and a plane
- use trigonometry to solve a variety of two dimensional and three dimensional problems.

A Trigonometry for any triangle

Our work on trigonometry so far has been concerned with right-angled triangles. We could deal with other types of triangles by drawing a perpendicular from one vertex to the opposite side, but this is usually a rather time-consuming and inefficient method.

We have formulae for the trig ratios of acute angles which occur in right-angled triangles. In non-right-angled triangles there may be an obtuse angle, and so we shall need to know how to find the trig ratios of an obtuse angle.

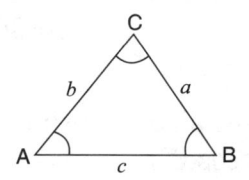

In trigonometrical formulae, capital letters refer to the angles of the triangle, and the small letters refer to the sides.

In a triangle ABC, we use A, B, C for the angles and a, b, c for the sides.
a is the side opposite angle A,
b is the side opposite angle B,
c is the side opposite angle C.

Sine and cosine of an obtuse angle

Easting and northing

Village Q is d kilometres from village P on a bearing of $A°$.

How far east and how far north is Q from P? Suppose A is an acute angle.

From the diagram, you can see that

$$\sin(A) = \frac{\text{opp}(A)}{\text{hypotenuse}} = \frac{RQ}{d}$$

and $\cos(A) = \dfrac{\text{adj}(A)}{\text{hypotenuse}} = \dfrac{PR}{d}$

so $RQ = d\sin(A)$ and $PR = d\cos(A)$.

RQ is the **easting** of Q from P and PR is the **northing**.

Hence, easting $= d\sin(A)$ and northing $= d\cos(A)$.

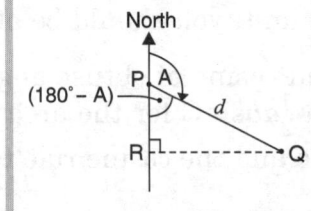

Now suppose A is an obtuse angle.

In the diagram, angle $QPR = 180° - A$, so, using the right-angled triangle QPR, we have

$$\sin(180° - A) = \frac{RQ}{d} \text{ and } \cos(180° - A) = \frac{PR}{d}.$$

Hence, $RQ = d\sin(180° - A)$
and $PR = d\cos(180° - A)$.

In this case, Q is *south* of P so we say the northing is negative. Q is still east of P, so the easting is positive.

Hence, easting $= d\sin(180° - A)$ and northing $= -d\cos(180° - A)$.

In order to have the same formulae for easting and northing for acute angle and obtuse angle bearings, we make the following definitions:

> If A is obtuse, $\sin(A) = +\sin(180° - A)$
> and $\cos(A) = -\cos(180° - A)$.

These definitions are programmed into all scientific calculators. Use your calculator to check that:
$\sin(100°) = +0.984807753 = \sin(80°)$
$\cos(100°) = -0.173648177 = -\cos(80°)$
$\sin(150°) = +0.5 \qquad\qquad = \sin(30°)$
$\cos(150°) = -0.866025403 = -\cos(30°)$

Your calculator will also give the results:
$\sin(0°) \;\; = 0 \qquad \cos(0°) \;\; = 1$
$\sin(90°) = 1 \qquad \cos(90°) = 0$
$\sin(180°) = 0 \qquad \cos(180°) = -1$

Can you explain these using the ideas of easting and northing?

Area of a triangle

How can you find the area of a triangle when you know the lengths of two sides and the size of the angle between these sides?

Suppose the known sides are a and b, and C is the angle between them.

a) If C is acute and side a is taken as the base, the perpendicular height (h) of the triangle is given by $\sin(C) = \frac{h}{b}$. Hence $h = b\sin(C)$.

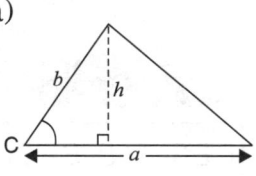

The area of the triangle
$= \frac{1}{2}$ base × perpendicular height
$= \frac{1}{2} a \times b\sin(C)$
$= \frac{1}{2} ab\sin(C)$

b) If C is obtuse, h is given by
$\sin(180° - C) = \frac{h}{b}$.
Hence $h = b\sin(180° - C)$.
The area of the triangle
$= \frac{1}{2}$ base × perpendicular height
$= \frac{1}{2} a \times b\sin(180° - C)$
$= \frac{1}{2} ab\sin(180° - C)$

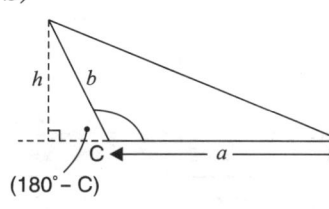

However, we have already defined the sine of an obtuse angle to be the same as the sine of (180° − the angle), so $\frac{1}{2} ab\sin(180° - C) = \frac{1}{2} ab\sin(C)$.

We can therefore use the following formula for all triangles:

$$\text{the area of a triangle} = \tfrac{1}{2} ab\sin(C)$$

Note the following:
1. Sin(C) is usually written as sinC, but the brackets in sin(180° − C) cannot be omitted.
2. Because $\sin 90° = 1$, the formula applies to right-angled triangles.
3. The formula may be written in words:
   Area of a triangle
   = half (product of any two sides) × (sine of the angle between them).
   The area of triangle ABC can be written as
   $\frac{1}{2} ab\sin C$ or $\frac{1}{2} bc\sin A$ or $\frac{1}{2} ca\sin B$.

Notice that, starting with the first version, the letters have moved round cyclically to obtain the other versions.

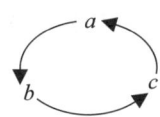

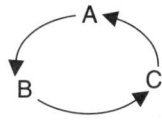

Similarly, area of a triangle PQR $= \frac{1}{2} pq\sin R = \frac{1}{2} qr\sin P = \frac{1}{2} rp\sin Q$.

Example 1

Use your calculator to find, correct to 4 decimal places:
a) sin 120°   b) cos 120°
c) sin 135°   d) cos 135°

Solution

a) Press the keys [1] [2] [0] [sin].

   The display is | DEG   0.8660 2540 3 |

   so sin(120°) = 0.8660 to 4 decimal places.

b) Press the keys [1] [2] [0] [cos].

   The display is | DEG   −0.5 |

   so cos(120°) = −0.5 exactly.

c) Press the keys [1] [3] [5] [sin].

   The display is | DEG   0.7071 0678 1 |

   so sin(135°) = 0.7071 to 4 decimal places.

d) Press the keys [1] [3] [5] [cos].

   The display is | DEG −0.7071 0678 1 |

   so cos(135°) = −0.7071 to 4 decimal places.

Example 2

Find the area of each of the shapes below.

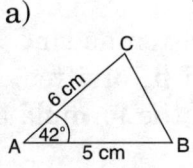

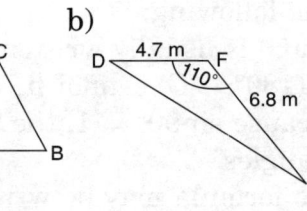

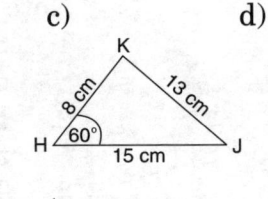

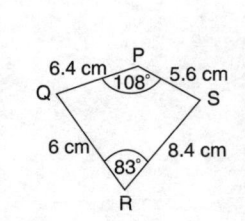

a) Triangle ABC with A = 42°, AB = 5 cm, AC = 6 cm.
b) Triangle DEF with DF = 4.7 m, FE = 6.8 m, angle F = 110°.
c) Triangle HJK with HK = 8 cm, KJ = 13 cm, HJ = 15 cm, angle H = 60°.
d) Quadrilateral PQRS with QP = 6.4 cm, PS = 5.6 cm, QR = 6 cm, RS = 8.4 cm, angle P = 108°, angle R = 83°.

Solution

a) $A = 42°$, $b = 6$ cm and $c = 5$ cm.
   Area of triangle $= \frac{1}{2}bc\sin A = \frac{1}{2} \times 6 \times 5 \times \sin 42°$
   $= 10.0369591$ [calculator display]
   $= 10.0$ cm² to 3 significant figures

b) $F = 110°$, $d = 6.8$ m and $e = 4.7$ m.
   Area of triangle $= \frac{1}{2}de\sin F = \frac{1}{2} \times 6.8 \times 4.7 \times \sin 110°$
   $= 15.01628808$ [calculator display]
   $= 15.0$ m² to 3 significant figures

c) $H = 60°$, $j = 8$ cm and $k = 15$ cm.
Area of triangle $= \frac{1}{2}jk\sin H = \frac{1}{2} \times 8 \times 15 \times \sin 60°$
$= 51.96152423$ [calculator display]
$= 52.0$ cm^2 to 3 significant figures

d) Divide the quadrilateral into two triangles, PQS and RQS.
Area of triangle PQS $= \frac{1}{2} \times 6.4 \times 5.6 \times \sin 108°$
$= 17.04293277$ cm^2

Area of triangle RQS $= \frac{1}{2} \times 6 \times 8.4 \times \sin 83°$
$= 25.01216302$ cm^2

Area of quadrilateral PQRS $= 42.05509579$ cm^2
$= 42.1$ cm^2 to 3 significant figures

Example 3

Find the area of an equilateral triangle in which each side is 8 cm.

Solution

Area of triangle $= \frac{1}{2}ab\sin C$
$= \frac{1}{2} \times 8 \times 8 \times \sin 60°$
$= 27.71281292$ cm^2
$= 27.7$ cm^2 to 3 significant figures

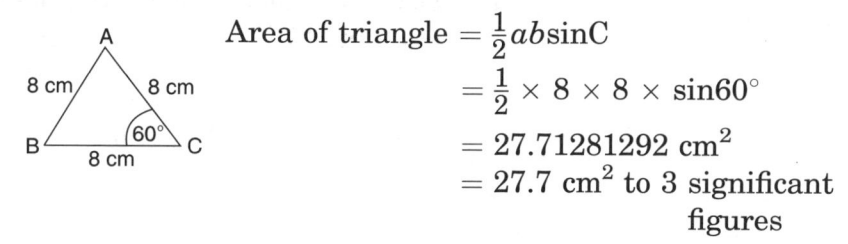

Example 4

In an acute-angled triangle DEF, the sine of angle D is 0.45. Find angle D correct to the nearest degree.

Solution

On the calculator, press the keys . 4 5 SHIFT sin^{-1} .

The display is DEG 26.74368395

Hence, angle D $= 27°$ to the nearest degree.

Example 5

In triangle ABC, side AB $= 6$ cm and side BC $= 10$ cm.
The area of the triangle is 15 cm^2.
What are the possible sizes of angle B?

Solution

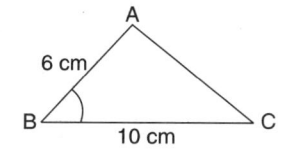

Area of triangle ABC $= \frac{1}{2}ca\sin B$
where $a = 10$ cm and $c = 6$ cm.

Hence, $\frac{1}{2} \times 6 \times 10 \times \sin B = 15$
$30 \times \sin B = 15$
$\sin B = 0.5$

On the calculator, press the keys [.] [5] [SHIFT] [sin⁻¹].

The display is | DEG   30 |

This means that angle B could be 30°.
However, sin150° = sin30° because 180° − 30° = 150°.
Hence, angle B is either 30° or 150°.

Scientific calculators are programmed to give the smallest possible value of the angle in cases such as this. You will have to decide whether there is more than one possible answer. The problem does not arise when you are working with cosines. The *cosine* of an *obtuse* angle is *negative*, whereas the cosine of an acute angle is positive.

Did you find it easy to follow this work on obtuse angles and the formula $\frac{1}{2}ab\sin C$ for the area of a triangle? See if you can answer these questions.

> You can use your calculator to find the *tangent* of an obtuse angle. (You will find that it is negative.) However, this is not in the IGCSE syllabus.

EXERCISE 25

1. Use your calculator to find, correct to 4 decimal places:
   a) sin145°
   b) cos145°
   c) sin150°
   d) cos150°

2. Find the area of each of the shapes below.

   a) 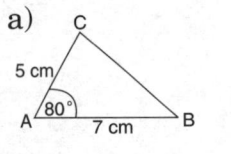 Triangle ABC with AC = 5 cm, AB = 7 cm, angle A = 80°.
   b) 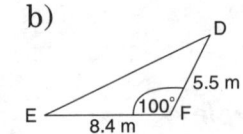 Triangle DEF with DF = 5.5 m, EF = 8.4 m, angle F = 100°.
   c) 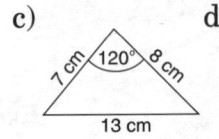 Triangle with sides 7 cm and 8 cm, included angle 120°, base 13 cm.
   d) 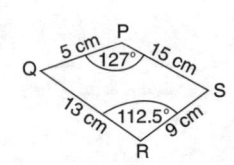 Quadrilateral PQRS with QP = 5 cm, PS = 15 cm, angle P = 127°, QR = 13 cm, RS = 9 cm, angle R = 112.5°.

3. 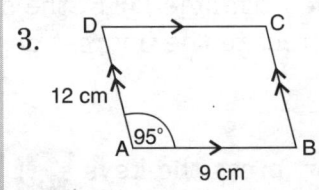 Find the area of the parallelogram ABCD in which AB = 9 cm, AD = 12 cm and angle A = 95°.

4. a) Given that sinA = 0.83, find the value of A to the nearest degree:
      (i) when A is an acute angle
      (ii) when A is an obtuse angle
   b) Find, correct to 1 decimal place, the obtuse angle which has a cosine of −0.48.

5. 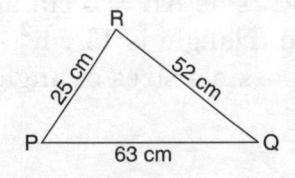 The diagram shows a triangle PQR which has an area of 630 cm². (PR = 25 cm, QR = 52 cm, PQ = 63 cm)
   a) Explain how you can tell that angle Q is an acute angle.
   b) Use the formula area = $\frac{1}{2}pr\sin Q$ to find angle Q correct to 1 decimal place.
   c) Find angle P correct to 1 decimal place.

Solving a triangle

Check your answers at the end of this module.

A triangle has three sides and three angles. To draw the triangle, you don't need to know the lengths of all the sides and the sizes of all the angles. You could draw the triangle if you knew:

- the lengths of all three sides
- the lengths of two sides and the size of the angle between them
- the length of one side and the sizes of two angles.

If you can *draw* the triangle, then it must be possible to *calculate* its remaining sides and angles. This is called **solving the triangle**.

The formulae we use for solving a triangle are called the **sine rule** and the **cosine rule**.

The sine rule

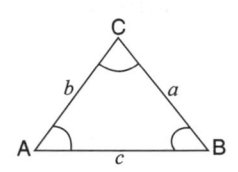

The area of triangle ABC can be written as

$\frac{1}{2}bc\sin A$ or $\frac{1}{2}ca\sin B$ or $\frac{1}{2}ab\sin C$.

It follows that

$\frac{1}{2}bc\sin A = \frac{1}{2}ca\sin B = \frac{1}{2}ab\sin C$.

Dividing by $\frac{1}{2}abc$ gives us

$\frac{\sin A}{a} = \frac{\sin B}{b} = \frac{\sin C}{c}$.

This is a version of the sine rule which is convenient for calculating angles. Inverting each fraction gives us the version which is more convenient for calculating sides:

$\frac{a}{\sin A} = \frac{b}{\sin B} = \frac{c}{\sin C}$

> both versions indicate that the sides of a triangle are proportional to the sines of the angles opposite to them

You need to remember:

> for any triangle ABC, $\frac{a}{\sin A} = \frac{b}{\sin B} = \frac{c}{\sin C}$
>
> and $\frac{\sin A}{a} = \frac{\sin B}{b} = \frac{\sin C}{c}$

Of course, you may have to use different letters in different triangles. For example, in triangle PQR,

$\frac{p}{\sin P} = \frac{q}{\sin Q} = \frac{r}{\sin R}$ and $\frac{\sin P}{p} = \frac{\sin Q}{q} = \frac{\sin R}{r}$.

To make use of the sine rule, you need to know the numerator and denominator of one of the fractions, and the numerator or denominator of one of the other fractions.

Example 1

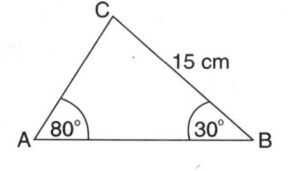

In triangle ABC, angle A = 80°, angle B = 30° and side BC = 15 cm.

Calculate the size of angle C and the lengths of the sides AB and AC.

Solution

The angle sum of the triangle is 180° so
  angle C = 180° − (80° + 30°)
  angle C = 70°

We now have to calculate the lengths of sides so we use the sine rule in the form

$$\frac{a}{\sin A} = \frac{b}{\sin B} = \frac{c}{\sin C}$$

Remembering that $a =$ BC, $b =$ AC and $c =$ AB, we have

$$\frac{15}{\sin 80°} = \frac{AC}{\sin 30°} = \frac{AB}{\sin 70°}$$

Using the first and second fractions, $AC = \frac{15 \times \sin 30°}{\sin 80°}$
  $= 7.615699589$

Using the first and third fractions, $AB = \frac{15 \times \sin 70°}{\sin 80°}$
  $= 14.31283341$

Hence, to 3 significant figures, AB = 14.3 cm and AC = 7.62 cm.

> as soon as you know the lengths of all the sides and the sizes of all angles, you should check that the shortest side is opposite the smallest angle, and the longest side is opposite the largest angle

Example 2

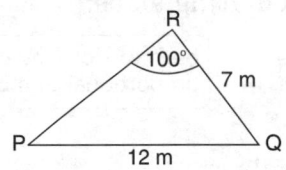

In triangle PQR, angle R = 100°, side PQ = 12 m and side QR = 7 m.

Calculate the sizes of angles P and Q, and the length of side PR.

Solution

We shall calculate angle P first, so we use the sine rule in the form

$$\frac{\sin P}{q} = \frac{\sin Q}{q} = \frac{\sin R}{r}$$

Remembering that $p =$ QR, $q =$ PR and $r =$ PQ, we have

$$\frac{\sin P}{7} = \frac{\sin Q}{q} = \frac{\sin 100°}{12}$$

Using the first and third fractions, $\sin P = \frac{7 \times \sin 100°}{12}$
  $= 0.574471189$

A triangle cannot have two obtuse angles, so angle P must be acute.
Hence, angle P = $\sin^{-1}(0.574471189)$
  $= 35.06260766°$
  angle P = 35.1° to 1 decimal place

The angle sum of the triangle is 180°, so
  angle Q = 180° − (100° + 35.0626°)
  $= 44.9374°$
  angle Q = 44.9° to 1 decimal place

Now using the sine rule in the form $\frac{q}{\sin Q} = \frac{r}{\sin R}$ we have

$$\frac{PR}{\sin 44.9374°} = \frac{12}{\sin 100°}$$

so $PR = \frac{12 \times \sin 44.9374°}{\sin 100°} = 8.6067\ldots$

side $PR = 8.61$ m to 3 significant figures

Example 3

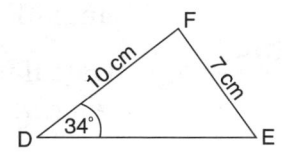

In triangle DEF, side DF = 10 cm, side EF = 7 cm and angle D = 34°.

Calculate, to the nearest degree, the possible sizes of angle E and angle F.

Solution

We shall calculate angle E first (because we know the length of the side opposite E).

$$\frac{\sin D}{d} = \frac{\sin E}{e} = \frac{\sin F}{f}$$

Remembering that $d = EF$, $e = DF$ and $f = DE$,

$$\frac{\sin 34°}{7} = \frac{\sin E}{10} = \frac{\sin F}{f}$$

Using the first and second fractions, $\sin E = \frac{10 \times \sin 34°}{7}$

$$= 0.798847005$$

E could be the acute angle $53.02014006°$
but it could be the obtuse angle $126.9798599°$
(There is no reason to reject either of these values.)

Angle F is found by using the angle sum of a triangle = 180°.

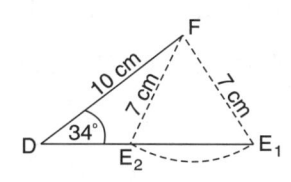

Working to the nearest degree, we obtain angle E = 53° and angle F = 93°
(this gives triangle DFE_1 in the diagram)
or angle E = 127° and angle F = 19°
(this gives triangle DFE_2 in the diagram).

You should now be able to use the sine rule to solve some triangles.

EXERCISE 26

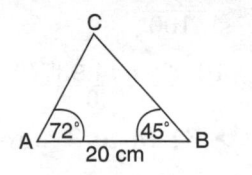

1. In triangle ABC, angle A = 72°, angle B = 45° and side AB = 20 cm.

   Calculate the size of angle C and the lengths of the sides AC and BC.

2. In triangle DEF, angle D = 140°, angle E = 15° and side DF = 6 m.

   Calculate the size of angle F and the lengths of the sides DE and EF.

3. In triangle PQR, angle Q = 120°, side PQ = 8 cm and side PR = 13 cm.

   Calculate the size of angle R, the size of angle P, and the length of side QR.

4. In triangle XYZ, angle X = 40°, side XZ = 12 cm and side YZ = 15 cm.
   a) Explain why angle Y must be less than 40°.
   b) Calculate, correct to 1 decimal place, angle Y and angle Z.
   c) Calculate the length of the side XY.

Check your answers at the end of this module. If you had any difficulty with any of the questions in Exercise 26, look carefully through the solutions and the worked examples on the previous pages.

The cosine rule

When all you know about a triangle is the lengths of the three sides, the sine rule is not useful for calculating the angles because, in $\frac{\sin A}{a} = \frac{\sin B}{b} = \frac{\sin C}{c}$ you know the denominators but not the value of any of the fractions.

Similarly, when you only know the lengths of two sides and the size of the included angle, you know the denominators of two of the fractions and the numerator of the third fraction, but you do not know the value of any of the fractions.

In these two cases, you will need to use another formula which is known as the ***cosine rule***.

The cosine rule is a generalisation of Pythagoras's theorem and it can be proved by using that theorem.

For the IGCSE examinations, you have to remember the cosine rule and be able to apply it correctly. You are not expected to be able to *prove* it, but the proof is given on the following page for your interest.

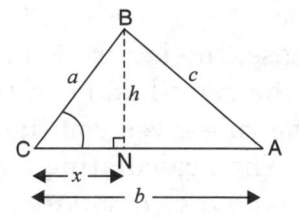

Consider the triangle ABC in the diagram, where angle C is acute.

The perpendicular, BN, from B to AC creates two right-angled triangles BNC and BNA.

Let $CN = x$ and $BN = h$.

In triangle BNC, $a^2 = h^2 + x^2$ — Pythagoras's theorem
In triangle BNA, $c^2 = h^2 + (b-x)^2$ — Pythagoras's theorem

Hence, $h^2 = a^2 - x^2$ and so $c^2 = a^2 - x^2 + (b-x)^2$
$= a^2 - x^2 + b^2 - 2bx + x^2$
$= a^2 + b^2 - 2bx$

We now use trigonometry in right-angled triangle BNC to obtain

$\cos C = \dfrac{\text{adjacent}}{\text{hypotenuse}} = \dfrac{x}{a}$ and hence $x = a\cos C$.

It follows that $c^2 = a^2 + b^2 - 2ab\cos C$.

This is the cosine rule. We have proved it in the case where C is acute. It also applies when C is a right-angle because $\cos 90° = 0$ and the rule becomes $c^2 = a^2 + b^2$, that is, Pythagoras's theorem.
We must now show that the cosine rule applies when C is obtuse.

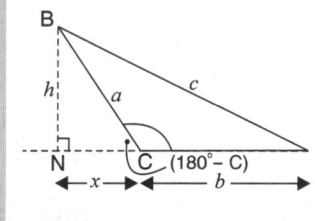

When angle C is obtuse, the right-angled triangles BNC and BNA overlap.
Applying Pythagoras's theorem, we obtain
$a^2 = h^2 + x^2$ and $c^2 = h^2 + (b+x)^2$

It follows that $c^2 = a^2 - x^2 + (b+x)^2$
$= a^2 - x^2 + b^2 + 2bx + x^2$
$= a^2 + b^2 + 2bx$

Using right-angled triangle BNC, angle $BCN = 180° - C$
and $\cos(BCN) = \dfrac{x}{a}$
so $x = a\cos(180° - C)$

We deduce that $c^2 = a^2 + b^2 + 2ab\cos(180° - C)$.

However, for any obtuse angle C, we have defined $\cos C$ to be equal to $-\cos(180° - C)$, so $c^2 = a^2 + b^2 - 2ab\cos C$.

We have thus established that the cosine rule applies when C is obtuse, and we can state:

> for any triangle ABC, $c^2 = a^2 + b^2 - 2ab\cos C$

Take note of the following:
1. To avoid the most common source of error in using the cosine rule, you should write it as $c^2 = a^2 + b^2 - (2ab\cos C)$.
   The values of a^2, b^2 and $(2ab\cos C)$ must be worked out before any addition or subtraction is done.
2. As with the formula for the area of a triangle, the letters can be moved round cyclically to obtain some other versions of the

cosine rule: $a^2 = b^2 + c^2 - (2bc\cos A)$ and $b^2 = c^2 + a^2 - (2ca\cos B)$. In each case, the letter at the left-hand end corresponds to the letter at the right-hand end (c and C, a and A, b and B).

3. The cosine rule gives you the *square* of one of the sides of the triangle. When calculating a side, you must remember that the last step is to *take a square root*.

Example 1

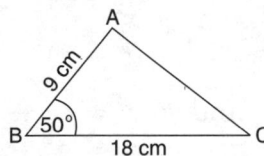

In triangle ABC, angle B = 50°, side AB = 9 cm and side BC = 18 cm.

Calculate the length of AC.

Solution

Notice that AC = b and we know that angle B = 50°.
We use the cosine rule in the form
$$b^2 = c^2 + a^2 - (2ca\cos B)$$
$$b^2 = 9^2 + 18^2 - (2 \times 9 \times 18 \times \cos 50°)$$
$$= 81 + 324 - (208.2631855)$$
$$= 196.7368145$$
$$b = \sqrt{196.7368145}$$
$$= 14.02629012$$

Length of AC = 14.0 cm to 3 significant figures.

Example 2

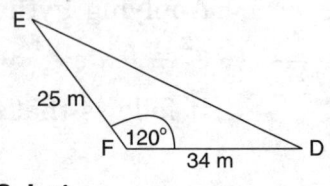

In triangle DEF, angle F = 120° side EF = 25 m and side FD = 34 m.

Calculate the length of side DE.

Solution

DE = f so we use the cosine rule in the form
$$f^2 = d^2 + e^2 - (2de\cos F)$$
$$f^2 = 25^2 + 34^2 - (2 \times 25 \times 34 \times \cos 120°)$$
$$= 625 + 1156 - (-850) \quad \boxed{\text{notice that } \cos 120° \text{ is negative}}$$
$$f^2 = 625 + 1156 + 850$$
$$f^2 = 2631$$
$$f = \sqrt{2631} = 51.29327441$$

Length of DE = 51.3 m to 3 significant figures.

Example 3

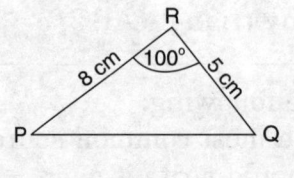

In triangle PQR, angle R = 100°, side PR = 8 cm and side RQ = 5 cm

a) Calculate the length of side PQ.
b) Calculate, correct to the nearest degree, angle P and angle R.

Solution

a) PQ = r so we use the cosine rule in the form
$$r^2 = p^2 + q^2 - (2pq\cos R)$$
$$r^2 = 5^2 + 8^2 - (2 \times 5 \times 8 \times \cos 100°)$$
$$r^2 = 25 + 64 - (-13.89185421)$$ [notice that cos100° is negative]
$$r^2 = 102.891854$$
$$r = \sqrt{102.8918542} = 10.14356221$$

Length of PQ = 10.1 cm to 3 significant figures.

b) Now we know the values of r and R, we can make use of the sine rule.
$$\frac{\sin P}{p} = \frac{\sin Q}{q} = \frac{\sin R}{r}$$
$$\frac{\sin P}{5} = \frac{\sin Q}{8} = \frac{\sin 100°}{10.14356\ldots}$$

Using the first and third fractions $\sin P = \frac{5 \times \sin 100°}{10.14356\ldots} = 0.485434866$

R is obtuse so P is acute, and angle P = 29.04096759°.

angle P = 29° to the nearest degree.

To find angle Q, we can use the angle sum of a triangle = 180°
angle Q = 180° − (100° + 29°)
so angle Q = 51° to the nearest degree.

Alternatively, we could go back to the sine rule statement and obtain
$$\sin Q = \frac{8 \times \sin 100°}{10.14356\ldots} = 0.776695955$$
This gives Q = 50.95904775°
so angle Q = 51° to the nearest degree.

Example 4

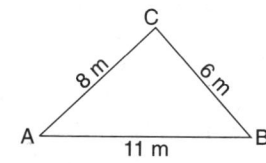

In triangle ABC, side AC = 8 m, side AB = 11 m and side BC 6 m.
a) Find the size of angle C.
b) Find the size of angle B.

Solution

a) We can use the cosine rule in the form
$$c^2 = a^2 + b^2 - (2ab\cos C) \text{ with } a = 6, b = 8 \text{ and } c = 11.$$
$$11^2 = 6^2 + 8^2 - (2 \times 6 \times 8 \times \cos C)$$
$$121 = 36 + 64 - (96\cos C)$$
$$96\cos C = 36 + 64 - 121$$
$$96\cos C = -21$$
$$\cos C = \frac{-21}{96}$$
$$c = \cos^{-1}\left(\frac{-21}{96}\right)$$
$$c = 102.6356251°$$
angle C = 102.6° correct to 1 decimal place

b) There is a choice of methods for finding angle B.
We could use the cosine rule again, in the form
$$b^2 = c^2 + a^2 - (2ca\cos B)$$
giving $8^2 = 11^2 + 6^2 - (2 \times 11 \times 6 \times \cos B)$
and $\cos B = \frac{93}{132}$

Alternatively, we could use the sine rule, in the form
$$\frac{\sin B}{b} = \frac{\sin C}{c}$$
giving $\frac{\sin B}{8} = \frac{\sin 102.635\ldots°}{11}$
and $\sin B = 0.70965\ldots$

The cosine rule method uses only the given information (the lengths of the three sides) and does not depend on the answer to part a), which may be incorrect.

However, many people think that the sine rule is easier to use than the cosine rule. You must decide for yourself!

Using either method, we obtain B = 45.207...°
and so angle B = 45.2° correct to 1 decimal place.

Example 5

a) Change the subject of the formula $c^2 = a^2 + b^2 - (2ab\cos C)$ to $\cos C$.
b) Use your answer to part a) to find the smallest angle in the triangle which has sides of length 7 m, 8 m and 13 m.

Solution

a)
$$c^2 = a^2 + b^2 - (2ab\cos C)$$
$$(2ab\cos C) = a^2 + b^2 - c^2$$
$$\cos C = \frac{a^2 + b^2 - c^2}{2ab}$$

b) The smallest angle in a triangle is opposite the shortest side. In the given triangle, the smallest angle is opposite the 7 m side. Let this angle be C. Then $c = 7$, and we can take $a = 8$ and $b = 13$. Using the result of part a),
$$\cos C = \frac{8^2 + 13^2 - 7^2}{2 \times 8 \times 13}$$
$$\cos C = \frac{64 + 169 - 49}{208}$$
$$\cos C = \frac{184}{208}$$
$$C = \cos^{-1}\left(\frac{184}{208}\right) = 27.7957725°$$

The smallest angle of the triangle = 27.8° to 1 decimal place.

When you know the lengths of all three sides of a triangle and you want to find the size of an angle, you can use the cosine rule in the form:

$$\cos C = \frac{a^2 + b^2 - c^2}{2ab}$$

As with the other version, the letters can be moved cyclically to give $\cos A = \dfrac{b^2 + c^2 - a^2}{2bc}$ and $\cos B = \dfrac{c^2 + a^2 - b^2}{2ca}$.

Now you should practise using the cosine rule and the sine rule by solving the following problems.

EXERCISE 27

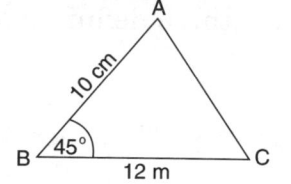

1. In triangle ABC, angle B = 45°, side AB = 10 cm and side BC = 12 cm.

   Calculate the length of side AC.

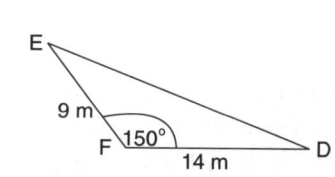

2. In triangle DEF, angle F = 150°, side EF = 9 m and side FD = 14 m.

   Calculate the length of side DE.

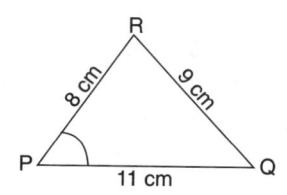

3. In triangle PQR, side PQ = 11 cm, side QR = 9 cm and side RP = 8 cm.

   Calculate the size of angle P correct to 1 decimal place.

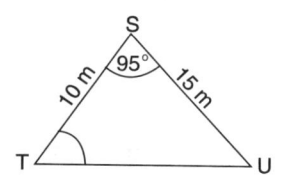

4. In triangle STU, angle S = 95°, side ST = 10 m and side SU = 15 m.

   a) Calculate the length of side TU.
   b) Calculate angle U.
   c) Calculate angle T.

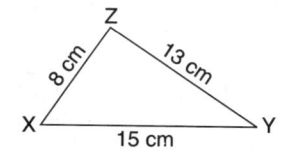

5. In triangle XYZ, side XY = 15 cm, side YZ = 13 cm and side ZX = 8 cm.

   Calculate the size of:

   a) angle X
   b) angle Y
   c) angle Z

Check your answers at the end of this module.

B The sine and cosine functions

So far, we have considered the sine and cosine of angles up to 180° because each angle in a triangle must be an acute angle, a right angle or an obtuse angle. There is no mathematical reason why we should stop at 180°. Angles greater than 180° occur when we consider rotating shafts in machinery, 3-figure bearings, etc. We could use 3-figure bearings to obtain useful definitions for the sine and cosine of reflex angles (as we did for obtuse angles in Section A).

You need not learn these definitions, but they are given here for reference:

> If $180° \leq A \leq 270°$, then $\sin A = -\sin(A - 180°)$
> and $\cos A = -\cos(A - 180°)$.
> If $270° < A \leq 360°$, then $\sin A = -\sin(360° - A)$
> and $\cos A = +\cos(360° - A)$.

You need not learn these definitions because the values of sinA and cosA are programmed into your calculator for all sizes of angle A. After A = 360°, the values of sinA and cosA repeat themselves. For example, sin370° is the same as sin10°, and cos370° is the same as cos10°. (A bearing of 370° is the same as a bearing of 010°.)

For the IGCSE examinations you have to be able to draw graphs of sine and cosine functions. To do this you need your scientific calculator. It will give values to 9 decimal places – far too many for drawing graphs! You will need to round these values. Usually 2 or 3 significant figures is sufficient. Some values will be positive but some may be negative. Be careful with the signs when you are reading the calculator display.

Example 1

Draw the graph of $y = \sin x°$ for $0 \leq x \leq 360$.

Solution

Using a calculator, and rounding values to 2 decimal places, we obtain this table.

x	0	30	60	90	120	150	180	210	240	270	300	330	360
$\sin x°$	0	0.50	0.87	1	0.87	0.50	0	−0.50	−0.87	−1	−0.87	−0.50	0

Here is the graph.

Example 2

Draw the graph of $y = \sin x° + \cos x°$ for $0 \leq x \leq 360$.

Solution

Using a calculator, and rounding values to 2 decimal places, we obtain this table.

x	0	15	30	45	60	75	90	105	120	135	150	165	180
y	1	1.22	1.37	1.41	1.37	1.22	1	0.71	0.37	0	−0.37	−0.71	−1

x	195	210	225	240	255	270	285	300	315	330	345	360
y	−1.22	−1.37	−1.41	−1.37	−1.22	−1	−0.71	−0.37	0	0.37	0.71	1

Here is the graph.

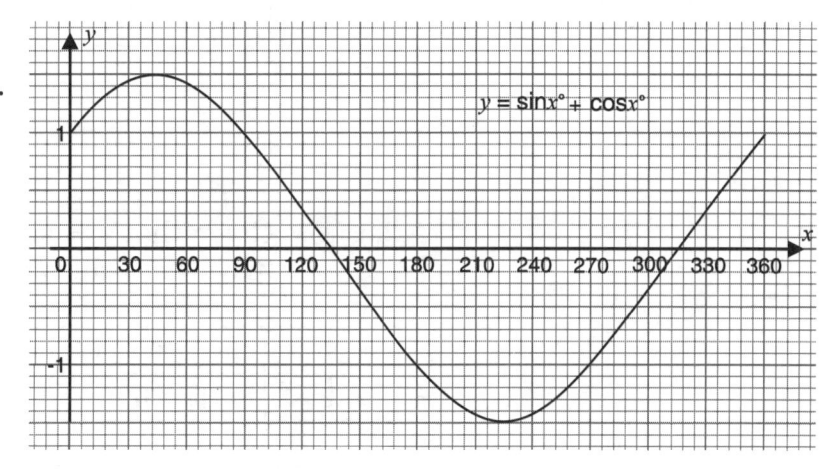

Example 3

a) Draw the graph of $y = 1 + 2\cos x°$ for $0 \leq x \leq 360$.
b) Use the graph to find two solutions of the equation $1 + 2\cos x° = 1.4$.

Solution

a) Here is a table of values, obtained by using a calculator.

x	0	30	60	90	120	150	180	210	240	270	300	330	360
y	3	2.73	2	1	0	−0.73	−1	−0.73	0	1	2	2.73	3

Here is the graph.

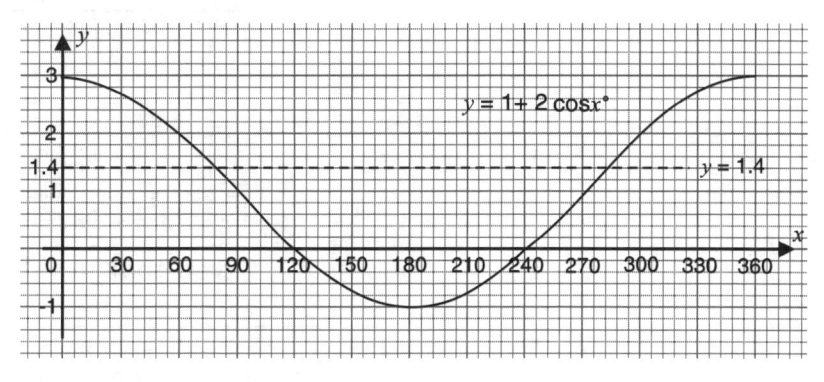

b) On the diagram, the line $y = 1.4$ has been drawn.
It intersects the graph of $y = 1 + 2\cos x°$ at the points where $x = 78$ and $x = 282$.

(each small square on the x-axis is 6 units)

Hence, two solutions of $1 + 2\cos x° = 1.4$ are $x = 78$ and $x = 282$.

Have you noticed any similarities between the graphs we have drawn? In Exercise 28 you are asked to draw some graphs for yourself. See if they have the same properties.

EXERCISE 28

1. On the axes below, draw the graph of $y = \cos x°$ for $0 \le x \le 360$. (Plot points for $x = 0, 30, 60, 90, \ldots, 360$.)

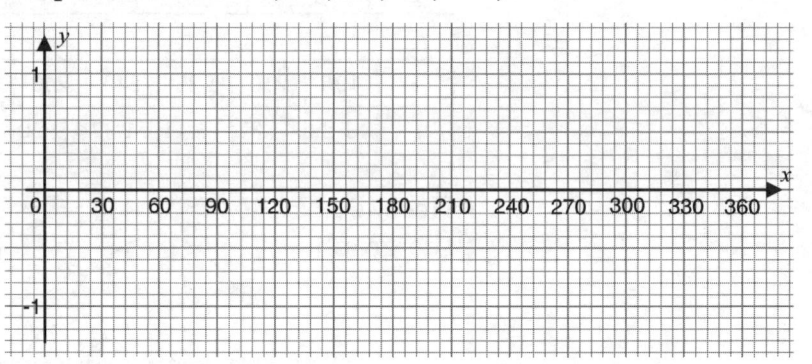

2. a) Given that $y = \sin x° - \cos x°$, complete this table of values. (Give the values correct to 2 decimal places.)

x	0	15	30	45	60	75	90	105	120	135	150	165	180
y	-1	-0.71	-0.37	0		0.71	1		1.37	1.41		1.22	1

x	195	210	225	240	255	270	285	300	315	330	345	360
y		0.37	0		-0.71	-1		-1.37	-1.41		-1.22	-1

   b) On the axes below, draw the graph of $y = \sin x° - \cos x°$ for $0 \le x \le 360$.

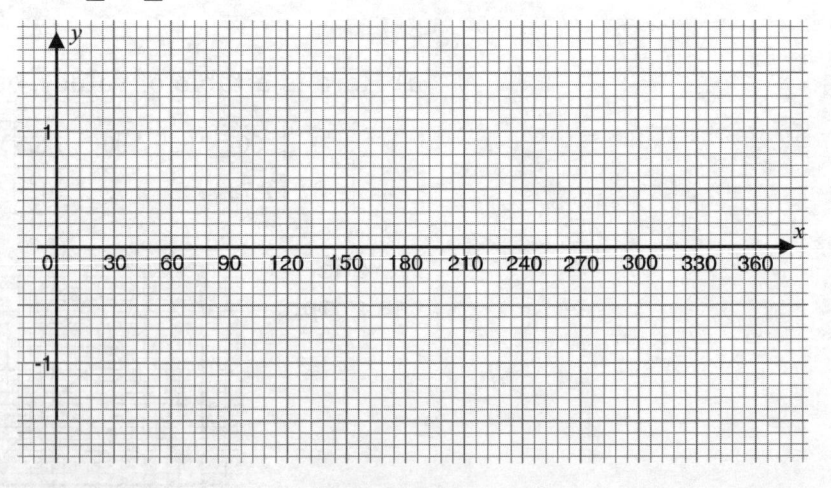

3. a) Given that $y = 1 + 2\sin x°$, complete this table of values.

x	0	30	60	90	120	150	180	210	240	270	300	330	360
y	1		2.73	3		2	1		-0.73	-1	-0.73		1

b) On the axes below, draw the graph of
$y = 1 + 2\sin x°$ for $0 \leq x \leq 360$.

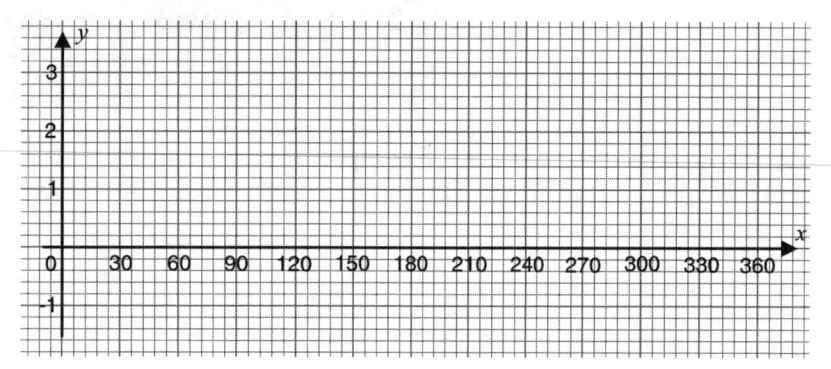

Can you now describe the common features of the graphs of sine and cosine functions which have been drawn? You will find the correct graphs at the end of this module.

C Further applications of trigonometry

The angle between a line and a plane

When you are dealing with solids, you may need to calculate the angle between an edge, or a diagonal, and one of the faces. This type of calculation also arises in other 3-dimensional problems. For example, to find the gradient of a path up a hillside.

These are particular cases of finding the angle between a line and a plane.

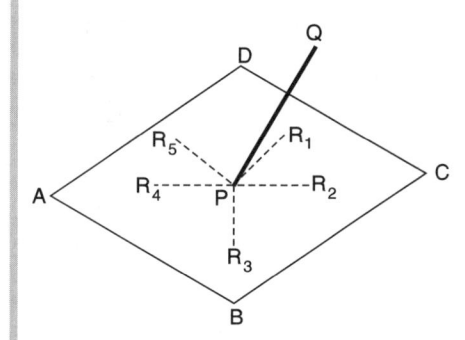

Consider a line PQ which meets a plane ABCD at the point P. Through P, draw lines PR_1, PR_2, PR_3, ... in the plane, and consider the angles QPR_1, QPR_2, QPR_3, ...

If PQ is perpendicular to the plane, all these angles will be right angles. If PQ is not perpendicular to the plane, these angles will vary in size.

It is the *smallest* of these angles which is called the angle between the line PQ and the plane ABCD.

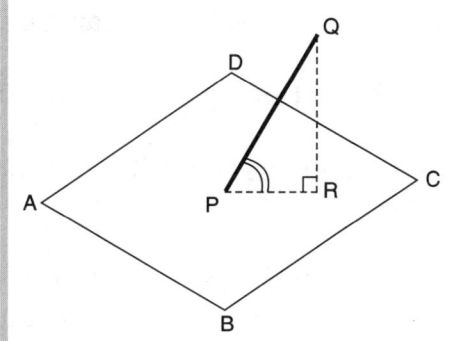

To identify this angle, we proceed as follows:
From the point Q, draw a perpendicular to the plane. Call the foot of this perpendicular R.
The angle between the line PQ and the plane ABCD is angle QPR.

PR is called the **projection** of PQ on the plane ABCD.

We can therefore say:

> The angle between a line and a plane is the angle between the line and its projection on the plane.

Example

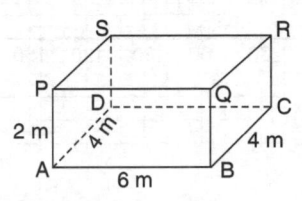

The diagram represents a room which has the shape of a cuboid. AB = 6 m, AD = 4 m and AP = 2 m.

Calculate the angle between the diagonal BS and the floor ABCD.

Solution

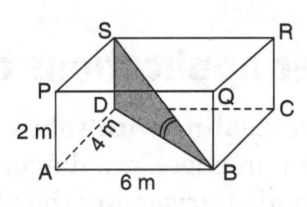

We must first identify the angle required. B is the point where the diagonal BS meets the plane ABCD.

SD is the perpendicular from S to the plane ABCD and so DB is the projection of SB on the plane.

The angle required is angle SBD.

We know that triangle SBD has a right angle at D and that SD = 2 m (equal to AP).

To find angle SBD, we need to know the length of BD or the length of SB.

We can find the length of BD by using Pythagoras's theorem in triangle ABD.

$$(BD)^2 = 6^2 + 4^2 = 36 + 16 = 52$$
$$BD = \sqrt{52}$$

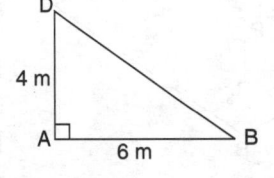

Hence, using right-angled triangle SBD,

$$\tan S\hat{B}D = \frac{\text{opposite}}{\text{adjacent}} = \frac{SD}{BD} = \frac{2}{\sqrt{52}}$$

$$\text{angle SBD} = \tan^{-1}\left(\frac{2}{\sqrt{52}}\right) = 15.50135957°$$

The angle between diagonal BS and the floor ABCD = 15.5° to 1 decimal place.

Using the formulae

To solve IGCSE problems, particularly the longer ones, you will need to use a variety of formulae and techniques. A problem which tests your understanding of trigonometry may also require an understanding of mensuration or geometry or (in some cases) algebra, and of technical terms such as 'bearing' and 'angle of depression'.

As far as trigonometry is concerned, you should know:

> the sine rule $\frac{a}{\sin A} = \frac{b}{\sin B} = \frac{c}{\sin C}$
>
> the cosine rule $c^2 = a^2 + b^2 - (2ab\cos C)$
>
> the area of a triangle $= \frac{1}{2}ab\sin C$

These apply to any triangles, including right-angled triangles (provided you remember that $\sin 90° = 1$ and $\cos 90° = 0$).

However, for right-angled triangles it is usually better to use:

> $\sin e = \frac{\text{opposite}}{\text{hypotenuse}}$, $\cos ine = \frac{\text{adjacent}}{\text{hypotenuse}}$, $\tan gent = \frac{\text{opposite}}{\text{adjacent}}$

Example 1

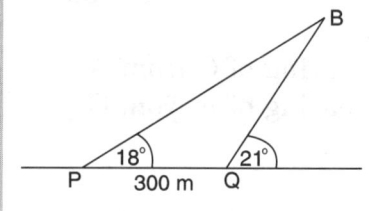

Two points, P and Q, are 300 m apart on horizontal ground. They are both due south of a stationary balloon, B.
From P, the angle of elevation of B is 18°, and from Q the angle of elevation is 21°.

Calculate the height of the balloon.

Solution

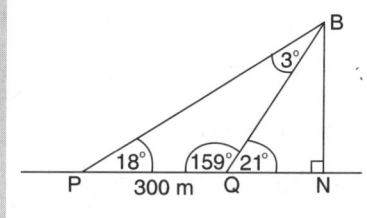

Let N be the point on the ground which is vertically below the balloon.
Angle PQB = 180° − 21° (angles on a
   = 159°   straight line)
angle PBQ = 21° − 18° (exterior angle
   = 3°   of triangle)

In triangle PQB, $\frac{b}{\sin B} = \frac{p}{\sin P} = \frac{q}{\sin Q}$

$$\frac{300}{\sin 3°} = \frac{BQ}{\sin 18°} = \frac{PB}{\sin 159°}$$

Hence, $BQ = \frac{300 \times \sin 18°}{\sin 3°} = 1771.346$

Triangle BQN is right-angled at N, so
$\sin 21° = \frac{\text{opposite}}{\text{hypotenuse}} = \frac{BN}{BQ}$

$BN = BQ \times \sin 21°$
$= 1771.346 \times \sin 21°$
$= 634.7937\ldots$

Height of the balloon = 635 m to 3 significant figures.

Example 2

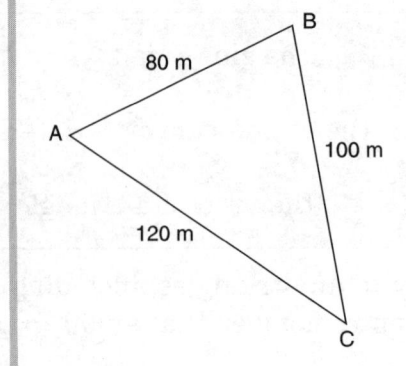

In a fitness exercise, students run across a field from A to B, then from B to C and then from C to A.

a) A student runs from A to B in 10 seconds. Calculate his speed in:
   (i) metres/second
   (ii) kilometres/hour
b) Another student runs from A to B in 10.5 seconds, from B to C in 13 seconds and from C to A at a speed of 8.5 m/s. Calculate her overall average speed in metres/second.
c) Showing all your working, calculate angle BAC.
d) The bearing of B from A is 062°.
Calculate:
   (i) the bearing of C from A
   (ii) the bearing of A from C

Solution

a) (i) Speed $= \dfrac{\text{distance}}{\text{time}} = \dfrac{80 \text{ m}}{10 \text{ s}} = 8$ m/s

   (ii) Speed $= 8$ m/s
   $= 8 \times 60 \times 60$ m/h *(there are 60 × 60 seconds in an hour)*
   $= \dfrac{8 \times 60 \times 60}{1000}$ km/h *(there are 1000 m in a km)*
   $= 28.8$ km/h

b) Time to run from C to A $= \dfrac{120 \text{ m}}{8.5 \text{ m/s}} = 14.117\ldots$ seconds

   Total distance $=$ AB $+$ BC $+$ CA $= 300$ m
   Total time $= (10.5 + 13 + 14.117\ldots)$ seconds $= 37.617\ldots$ seconds

   Overall average speed $= \dfrac{300 \text{ m}}{37.617\ldots \text{ seconds}} = 7.97498\ldots$ m/s
   $= 7.97$ m/s to 3 significant figures

c) To calculate angle BAC, we use the cosine rule in the form
   $$\cos A = \dfrac{b^2 + c^2 - a^2}{2bc}$$
   $$\cos A = \dfrac{120^2 + 80^2 - 100^2}{2 \times 120 \times 180} = \dfrac{10\,800}{19\,200} = \dfrac{9}{16}$$
   $$A = \cos^{-1}\left(\dfrac{9}{16}\right) = 55.7711\ldots °$$
   angle BAC $= 55.8°$ correct to 1 decimal place

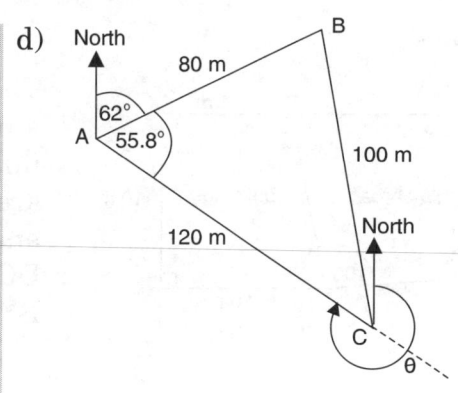

d)

(i) Bearing of C from A
$= 62° + 55.8°$
$= 117.8°$
$= 118°$ to the nearest degree

(ii) Bearing of A from C
$=$ angle θ in the diagram
$= 180° + 117.8°$
$= 297.8°$
$= 298°$ to the nearest degree

Example 3

The number of hours of daylight in the north of Iceland is given approximately by $12 - 12\cos(30t + 10)°$, where t is the time in months which has passed since 1st January.
a) Calculate the number of hours of daylight on 1st August.
b) Draw a graph of the function $h = 12 - 12\cos(30t + 10)°$ for $0 \leq t \leq 12$.
c) When will there be 11 hours of daylight?

Solution

a) 1st January to 1st August is 7 months.
When $t = 7$, $12 - 12\cos(30t + 10)° = 12 - 12\cos 220°$
$= 12 - (-9.1925\ldots)$
$= 21.1925\ldots$

Number of hours of daylight on 1st August $= 21.2$ to 3 significant figures.

b) This table gives values of h correct to 1 decimal place (obtained by using a calculator).

t	0	1	2	3	4	5	6	7	8	9	10	11	12
h	0.2	2.8	7.9	14.1	19.7	23.3	23.8	21.2	16.1	9.9	4.3	0.7	0.2

Here is the graph.

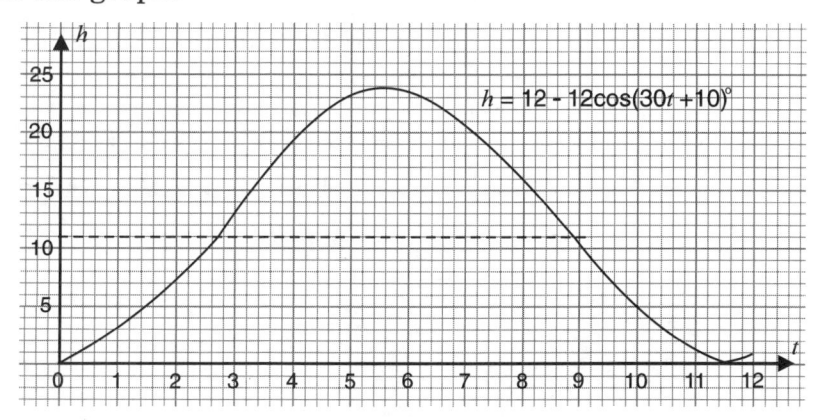

c) From the graph, $h = 11$ when $t = 2.7$ and when $t = 8.9$.
There will be (approximately) 11 hours of daylight on 21st March (2.7 months after 1st January) and on 27th September (8.9 months after 1st January).

Example 4

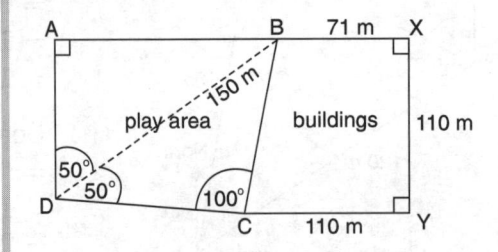

The diagram shows the plan for a new school. ABX is a straight line. The angles at A, X and Y are each 90°, angle ADB = 50°, angle BDC = 50° and angle DCB = 100°. DB = 150 m, BX = 71 m and XY = CY = 110 m.

Calculate, correct to 3 significant figures:
 (i) the area of BXYC
 (ii) the length of AD
 (iii) the length of AB
 (iv) the length of BC

b) Using your answers to part a), calculate the total area of the school grounds, AXYCD. Give your answer correct to 2 significant figures.

Solution

a) (i) BXYC is a trapezium with BX and CY as its parallel sides.
Area of BXYC = (average of parallel sides) ×
     (perpendicular distance between them)

$$= \left(\frac{71 + 110}{2}\right) \times 110 \text{ m}^2$$

$$= 9955 \text{ m}^2$$

Area of BXYC = 9960 m² to 3 significant figures.

(ii) Triangle ABD has a right angle at A

so $\cos 50° = \frac{\text{adjacent}}{\text{hypotenuse}} = \frac{AD}{150}$

AD = 150 × cos50°
  = 96.418 ...

Length of AD = 96.4 m to 3 significant figures.

(iii) In triangle ABD, $\sin 50° = \frac{\text{opposite}}{\text{hypotenuse}} = \frac{AB}{150}$

AB = 150 × sin50°
  = 114.9066

Length of AB = 115 m to 3 significant figures.

(iv) Triangle BCD is *not* right-angled.
We use the sine rule to find the length of BC.
In triangle BCD, angle B = 180° − (100° + 50°) = 30°
and
$$\frac{b}{\sin B} = \frac{c}{\sin C} = \frac{d}{\sin D}$$
$$\frac{CD}{\sin 30°} = \frac{150}{\sin 100°} = \frac{BC}{\sin 50°}$$

Hence $BC = \frac{150 \times \sin 50°}{\sin 100°} = 116.679\ldots$

Length of BC = 117 m to 3 significant figures.

b) Area of triangle ABD = $\frac{1}{2}$AB × AD = $\frac{1}{2}$(114.9066...) × (96.418...)
= 5539.54... m²

In △BCD, angle B = 180° − (100° + 50°) = 30° (angle sum of triangle)

Area of triangle BCD = $\frac{1}{2}cd\sin B$ = $\frac{1}{2}$(150 × 116.679...) × sin 30°
= 4375.47... m²

Area of trapezium BXYC = 9955 m²

Total area of school grounds = (5539.54... + 4375.47... + 9955) m²
= 19 870 m²

Total area of school grounds = 20 000 m² to 2 significant figures.

Here are four questions for you to try. You will find that they are similar to the examples I have just given you.

EXERCISE 29

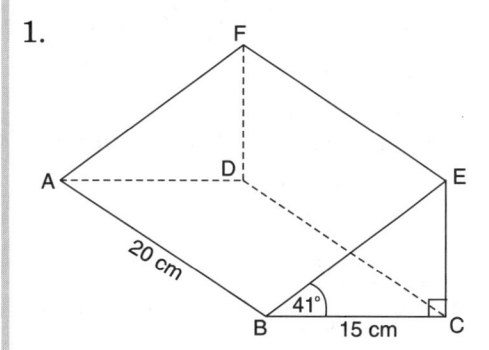

1. The diagram represents a triangular prism. The rectangular base, ABCD, is horizontal.
AB = 20 cm and BC = 15 cm.

The cross-section of the prism, BCE, is right-angled at C and angle EBC = 41°.
   a) Calculate the length of AC.
   b) Calculate the length of EC.
   c) Calculate the angle which the line AE makes with the horizontal.

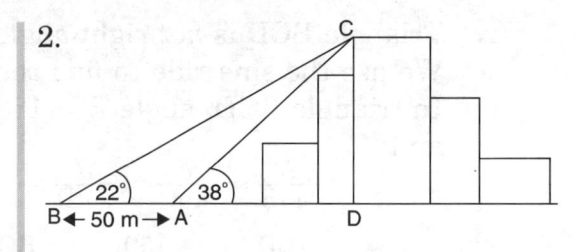

2. To find the height, CD, of a building, Jim took measurements from two points, A and B, 50 m apart, as shown in the diagram. B, A and D are in a straight line on horizontal ground and angle ADC = 90°.

The angle of elevation of the top of the building from A is 38°, and from B is 22°.
   a)   Calculate the distance BC.
   b)   Calculate the height, CD, of the building.

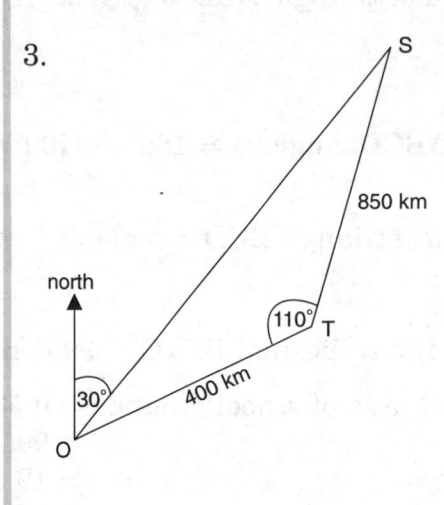

3. The diagram shows the relative positions of Osaka (O), Tokyo (T) and Sapporo (S) in Japan.
ST = 850 km, TO = 400 km and angle STO = 110°.
   a)   Calculate OS, the distance from Osaka to Sapporo.
   b)   Calculate the angle SOT, to the nearest degree.
   c)   The bearing of Sapporo from Osaka is 030°. Find the bearing of Osaka from Tokyo.
   d)   A plane flew from Sapporo to Tokyo at an average speed of 500 km/h. It left Sapporo at 0930. At what time did it arrive in Tokyo?

4. The depth, d metres, of water in a harbour on a certain day was given by $d = 4 - 3\sin(30t)°$, where t is the number of hours after 12 midnight.
   a)   Calculate the depth of water in the harbour at:
       (i) 2 a.m.
       (ii) 10 a.m.
   b)   Complete this table, showing values of d correct to 1 decimal place.

t	0	1	2	3	4	5	6	7	8	9	10	11	12
d	4	2.5		1		2.5	4		6.6	7		5.5	4

c) On the axes below, draw a graph of the function $d = 4 - 3\sin(30t)°$.

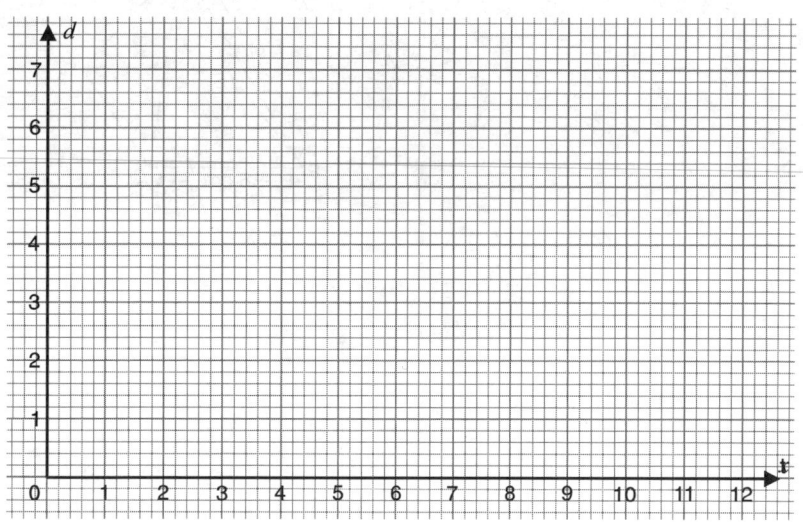

d) Between which times was the depth of water more than 6 m?

You may have found Exercise 29 quite difficult. Check your answers at the end of this module. If most of your answers are correct, you deserve congratulations.

Summary

You need to remember the following formulae to be able to solve the type of problems you've had in this unit:

- the area of a triangle $= \frac{1}{2}ab\sin C$
- the sine rule is $\frac{a}{\sin A} = \frac{b}{\sin B} = \frac{c}{\sin C}$ or $\frac{\sin A}{a} = \frac{\sin B}{b} = \frac{\sin C}{c}$
- the cosine rule is $c^2 = a^2 + b^2 - 2ab\cos C$.

Remember also that if A is obtuse, $\sin A = \sin(180° - A)$ and $\cos A = -\cos(180° - A)$.

In this unit you also saw how to draw graphs involving the sine and cosine functions by using your calculator to help you plot points. Your final task in Module 5 is to answer the questions in the 'Check your progress'. These questions have been taken from examination papers.

Check your progress

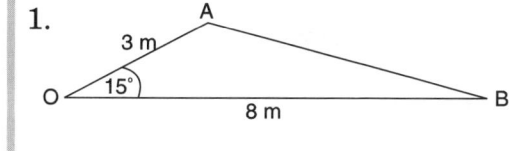

1. In the triangle OAB, angle AOB $= 15°$, OA $= 3$ m and OB $= 8$ m.

   Calculate, correct to 2 decimal places:
   a) the length of AB
   b) the area of triangle OAB

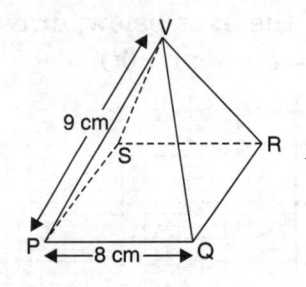

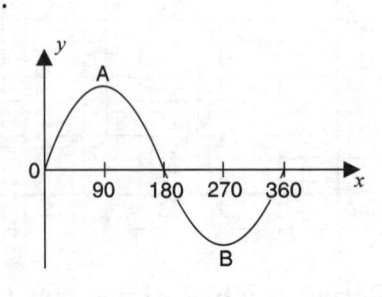

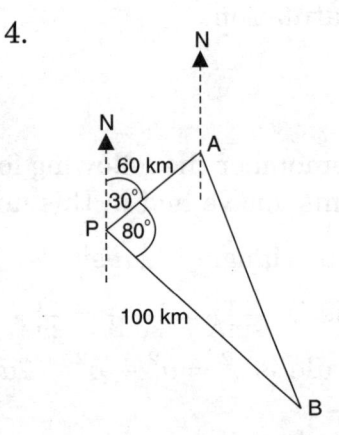

2. A pyramid, VPQRS, has a square base, PQRS, with sides of length 8 cm. Each sloping edge is 9 cm long.
   a) Calculate the perpendicular height of the pyramid.
   b) Calculate the angle which the sloping edge VP makes with the base.

3. The diagram shows the graph of $y = \sin x°$ for $0 \leq x \leq 360$.
   a) Write down the coordinates of A, the point on the graph where $x = 90$.
   b) Find the value of $\sin 270°$.
   c) Draw, on the diagram, the line $y = -\frac{1}{2}$ for $0 \leq x \leq 360$.
   d) How many solutions are there for the equation $\sin x = -\frac{1}{2}$, for $0 \leq x \leq 360$?

4. Two ships leave port P at the same time. One ship sails 60 km on a bearing of 030° to position A. The other ship sails 100 km on a bearing of 110° to position B.
   a) Calculate:
      (i) the distance AB
      (ii) angle PAB
      (iii) the bearing of B from A
   b) Both ships took the same time, t hours, to reach their positions. The speed of the *faster* ship was 20 km/h.
      Write down:
      (i) the value of t
      (ii) the speed of the slower ship

Check your answers at the end of this module.

Well done! You have now completed Module 5.

Solutions

EXERCISE 1

1. a) 48 mm × 4 = 192 mm
    b) 9 cm + 6 cm + 9 cm + 6 cm = 30 cm
    c) 5.2 cm + 5.2 cm + 5.2 cm = 15.6 cm
    d) 300 cm + 95 cm + 270 cm + 85 cm
        = 750 cm (or 7.5 m)

2. a) (2 + 1 + 1 + 2 + 1 + 2 + 2 + 2 + 1 + 2 + 1 + 1) cm
        = 18 cm   *same as for rectangle 4 cm by 5 cm*
    b) (2 + 3 + 1 + 1 + 2 + 1 + 1 + 1 + 2 + 2) cm
        = 16 cm   *same as for a square 4 cm by 4 cm*
    c) (3 + 5 + 3 + 1 + 2 + 1 + 1 + 1 + 1 + 1 + 2 + 1)
        = 22 cm

3. a) (25 + 18 + 23 + 20 + 16) m = 102 m
    b) (21 + 6 + 15 + 18 + 21 + 6 + 15 + 18) cm = 120 cm
    c) (6 + 8 + 4 + 2 + 10 + 2 + 4 + 6 + 2 + 2 + 2 + 4) cm = 52 cm

4. a) P = 3L
    b) P = S + S + B or P = 2S + B

5. There are 5 lines of length 68 m, 2 lines of length 140 m and 2 lines of length 94 m.
    Total length of the markings
    = (340 + 280 + 188) m = 808 m.

6. a) (85 + 35 + 85 + 35) cm = 240 cm
    b) (4.27 + 4.27 + 5.83 + 5.83) m = 20.20 m
    c) (338 + 338 + 338 + 224) mm = 1238 mm

EXERCISE 2

1. a) (9.5 × 7.4) cm² = 70.3 cm²
    b) $(3\frac{1}{2} \times 2\frac{2}{5})$ m² = $\frac{7}{2} \times \frac{12}{5}$ m² = $\frac{84}{10}$ m² = 8.4 m²
    c) (5.6 × 0.75) m² = 4.2 m²

2. (8.6 × 8.6) cm² = 73.96 cm²

3. a) (42.5 × 22.0) cm² = 935 cm²
    b) (135 × 86) cm² = 11 610 cm²

4. 1 km = 1000 m so
    1 km² = 1000 m × 1000 m = 1 000 000 m²
    Area of the field = 0.00357 × 1 000 000 m² = 3570 m²

5. Area = (125 × 90) m² = 11 250 m²
    1 hectare = 10 000 m²
    so area of the field = 1.125 hectares

EXERCISE 3

1. a) $\frac{1}{2}$(64 × 25) mm² = 800 mm²
    b) $\frac{1}{2}$(8 × 7) cm² = 28 cm²
    c) (5 × 2.4) m² = 12 m²
    d) $\frac{10+15}{2}$ × 8 m² = 100 m²

2. a) base = 3 cm, height = 2 cm
        so area = $\frac{1}{2}$(3 × 2) cm² = 3 cm²
    b) Take the left-hand side as the base.
        base = 3 cm, perpendicular height = $2\frac{1}{2}$ cm
        so area = (3 × $2\frac{1}{2}$) cm² = $7\frac{1}{2}$ cm²
    c) base = 3 cm   top side = 2 cm
        perpendicular height = 3 cm
        area = $\frac{3+2}{2}$ × 3 cm² = $7\frac{1}{2}$ cm²
    d) base = 1.5 cm, perpendicular height = 2.5 cm
        area = $\frac{1}{2}$(1.5 × 2.5) cm² = 1.875 cm²

3. a) $\frac{1}{2}$(8 × 3) cm² = 12 cm²
    b) $\frac{1}{2}$(12 × 5) m² = 30 m²
    c) (9 × 4) cm² = 36 cm²
    d) $\frac{9+6}{2}$ × 4 m² = 30 m²

EXERCISE 4

(Note: The answers to this test, and to later tests, have been obtained by using 3.142 as the value for π. If you use the π key on your calculator, you will obtain slightly different answers. In general, the difference will not be more than 1 in the third figure of the 3 significant figure answer. Where a calculator has been used, the calculator display is given for intermediate results. This is to enable you to check your work. As a general rule, you should show at least 4 significant figures in your working, unless the quantity is exact.)

1. a) 2π × 20 m = 125.68 m = 126 m to 3 significant figures
    b) 2π × 12.3 cm = 77.2932 cm
        = 77.3 cm to 3 significant figures
    c) π × 57 mm = 179.094 mm
        = 179 mm to 3 significant figures

2. Length of equator = 2π × 6400 km = 40 217.6 km
    = 40 200 km to 3 significant figures.

3. Length of cotton = (π × 2.3 cm) × 900
    = 6503.94 cm = 65.0 m to 3 significant figures.

EXERCISE 5

1. 28 cm^2

2. a) $\pi(5.2)^2 \text{ cm}^2 = 84.95968 \text{ cm}^2$
      $= 85.0 \text{ cm}^2$ to 3 significant figures
   b) $\pi(0.78)^2 \text{ m}^2 = 1.9115928 \text{ m}^2$
      $= 1.91 \text{ m}^2$ to 3 significant figures
   c) $\pi(23)^2 \text{ mm}^2 = 1662.118 \text{ mm}^2 = 1660 \text{ m}^2$ to 3 significant figures

3. $\pi(7.5)^2 \text{ m}^2 = 176.7375 \text{ m}^2$
   $= 177 \text{ m}^2$ to 3 significant figures

4. Area to be sealed $= \pi(5.2)^2 \text{ m}^2 = 84.95968 \text{ m}^2$
   $84.95968 \div 9 = 9.439964444$
   so 10 drums of sealant will have to be bought.

EXERCISE 6

1. a) Area $= ((8 \times 2) + (5 \times 2) + (8 \times 2)) \text{ cm}^2$ or $((8 \times 9) - (3 \times 5) - (3 \times 5)) \text{ cm}^2 = 42 \text{ cm}^2$
   b) Area $= ((8 \times 10) + \frac{1}{2}(8 \times 6)) \text{ cm}^2 = 104 \text{ cm}^2$
   c) Area $= ((12 \times 11) - \frac{1}{2}(6 \times 8)) \text{ cm}^2 = 108 \text{ cm}^2$
   d) Area $= ((7 \times 2) + (3 \times 3) + (7 \times 2)) \text{ cm}^2$ or $((7 \times 7) - \frac{1}{2}(4 \times 3) - \frac{1}{2}(4 \times 3)) \text{ cm}^2 = 37 \text{ cm}^2$

2. Area $= ((8 \times 3) + (3 \times 4) + \frac{1}{2}\pi(4)^2) \text{ cm}^2 = (36 + 8\pi) \text{ cm}^2 = 61.136 \text{ cm}^2 = 61.1 \text{ cm}^2$ to 3 significant figures

3. a) Coordinates of C $= (-4, 4)$
   b) Using the 'box' method
      area of triangle ABC = area of rectangle − areas of 3 right-angled triangles
      $= ((12 \times 6) - \frac{1}{2}(4 \times 12) - \frac{1}{2}(6 \times 4) - \frac{1}{2}(8 \times 2)) \text{ units}^2$
      $= 28 \text{ units}^2$

4. Area of card remaining $= ((8 \times 15) - \pi(4)^2 - \pi(3)^2) \text{ cm}^2 = (120 - 25\pi) \text{ cm}^2$
   $= (120 - 78.55) \text{ cm}^2 = 41.45 \text{ cm}^2$
   $= 41.5 \text{ cm}^2$ to 3 significant figures

5. a) Area P $= (2 \times 2) \text{ m}^2 = 4 \text{ m}^2$ (square)   scale 1 cm : 1 m
      Area Q $= (8 \times 2) \text{ m}^2 = 16 \text{ m}^2$ (rectangle)
      Area R $= \frac{1}{2}(2 \times 2) \text{ m}^2 = 2 \text{ m}^2$ (triangle)
      Area S $= (5 \times 2) \text{ m}^2 = 10 \text{ m}^2$ (parallelogram)
      Area T $= \left(\frac{5+3}{2}\right) \times 2 \text{ m}^2 = 8 \text{ m}^2$ (trapezium)
   b) (i) Blue area $= 16 \text{ cm}^2$ Total area $= 40 \text{ cm}^2$ Fraction painted blue $= \frac{2}{5}$
      (ii) Red area $= (4 + 8) \text{ cm}^2 = 12 \text{ cm}^2$
           Percentage of total area painted red $= \frac{12}{40} \times 100\% = 30\%$

EXERCISE 7

1. Area of parallelogram = side AB × perpendicular distance between AB and DC
   $28.7 = 8.2 \times$ perpendicular distance between AB and DC
   Perpendicular distance between AB and DC $= \frac{28.7}{8.2} \text{ cm} = 3.5 \text{ cm}$

2. $\pi r^2 = 69.4$ so $r^2 = \frac{69.4}{3.142} = 22.08784214$ and $r = 4.699770435$
   Radius of the circle $= 4.7 \text{ cm}$ to 2 significant figures.

3. a) (i) Perimeter $= (5 + 2 + 5 + 2) \text{ m} = 14 \text{ m}$
      (ii) Side of square $= \frac{14}{4} \text{ m} = 3.5 \text{ m}$
           Area of square $= (3.5)^2 \text{ m}^2 = 12.25 \text{ m}^2$
   b) (i) Area of rectangle $= (5 \times 2) \text{ m}^2 = 10 \text{ m}^2$
      (ii) Side of square $= \sqrt{10} \text{ m} = 3.16227766 \text{ m}$
           $= 3.16 \text{ m}$ to 3 significant figures

EXERCISE 7 (cont.)

4. a) Circumference = $\pi \times$ diameter = 3.142×30 cm = 94.26 cm
    $= 94.3$ cm to 3 significant figures

    b) Circumference = $\pi \times$ diameter so diameter = $\frac{\text{circumference}}{\pi}$

    Diameter = $\frac{100}{3.142}$ cm = 31.82686187 cm = 32 cm to the nearest centimetre

5. a) Area of triangle = $\frac{1}{2}(9 \times 12)$ cm² = 54 cm²

    b) Area of triangle = $\frac{1}{2}(15 \times AN)$ so AN = $\frac{2 \times 54}{15}$ cm = 7.2 cm

EXERCISE 8

1. a) Each side of triangle ABC is $\frac{5}{3}$ times the corresponding side of triangle DEF. Since the ratio of corresponding sides is constant the triangles are similar.

    b) $\frac{\text{perimeter of triangle ABC}}{\text{perimeter of triangle DEF}} = \frac{15+20+25}{9+12\ 15} = \frac{60}{36} = \frac{5}{3}$

    c) The linear scale factor = $\frac{5}{3}$

    so the area scale factor = $\left(\frac{5}{3}\right)^2 = \frac{25}{9}$

    Hence, $\frac{\text{area of triangle ABC}}{\text{area of triangle DEF}} = \frac{25}{9}$

2. a) Each side of rectangle JKLM is $\frac{5}{3}$ times the corresponding side of rectangle PQRS, and the corresponding angles are equal (each angle is 90°) so the rectangles are similar.

    b) $\frac{\text{perimeter of rectangle JKLM}}{\text{perimeter of rectangle PQRS}} = \frac{10+15+10+15}{6+9+6+9}$

    $= \frac{50}{30} = \frac{5}{3}$

    c) $\frac{\text{area of rectangle JKLM}}{\text{area of rectangle PQRS}} = \frac{10 \times 15}{6 \times 9} = \frac{150}{54} = \frac{25}{9}$

3. a) Linear scale factor = $\frac{EF}{AB} = \frac{45}{25} = \frac{9}{5}$

    $\frac{EG}{AC}$ = ratio of corresponding lengths

    = linear scale factor = $\frac{9}{5}$

    b) $\frac{\text{perimeter of rhombus EFGH}}{\text{perimeter of rhombus ABCD}}$ = linear scale factor

    $= \frac{9}{5}$

    c) $\frac{\text{area of rhombus EFGH}}{\text{area of rhombus ABCD}}$ = area scale factor

    $= \left(\frac{9}{5}\right)^2 = \frac{81}{25}$

4. Area scale factor = $\frac{128}{72} = \frac{16}{9}$

    so linear scale factor = $\sqrt{\frac{16}{9}} = \frac{4}{3}$

    Longest side of second quadrilateral = $\frac{4}{3} \times 9$ cm

    $= 12$ cm

5. Linear scale factor = $10\,000 = 10^4$
    so area scale factor = $(10^4)^2 = 10^8$
    Actual area of the forest = 8×10^8 cm²
    1 square metre = 100 cm $\times 100$ cm
    $= 10\,000$ cm² = 10^4 cm²

    Hence, actual area of the forest
    $= \frac{8 \times 10^8}{10^4}$ m² = 8×10^4 m²

    Actual area of the forest = $80\,000$ m².

6. Triangle ABC is an enlargement of triangle AHG with linear scale factor 3.

    The area scale factor = $3^2 = 9$

    so area of triangle AHG = $\frac{1}{9}$ area of triangle ABC.

    Similarly, area of triangle IBD = $\frac{1}{9}$ of triangle ABC

    and area triangle FEC = $\frac{1}{9}$ area of triangle ABC.

    Hence, area of hexagon DEFGHI

    $= (1 - \frac{3}{9})$ area of triangle ABC

    $= \frac{2}{3}$ area of triangle ABC

EXERCISE 9

1. a) Arc length $= \frac{50}{360} \times (2 \times \pi \times 12)$ cm
        $= 10.47333333$ cm

    Perimeter of sector $= 10.47333333 + 12 + 12)$ cm
    $= 34.47333333$ cm
    $= 34.5$ cm to 3 significant figures

    b) Area of sector $= \frac{50}{360} \times \pi \times (12)^2 \text{cm}^2 = 62.84 \text{ cm}^2$
       $= 62.8 \text{ cm}^2$ to 3 significant figures

2. Area wiped $= \frac{120}{360} \times \pi \times (15+55)^2 - \frac{120}{360} \times \pi \times (15)^2 \text{cm}^2$
    $= \frac{120}{360} \times \pi \times (70^2 - 15^2) \text{ cm}^2$
    $= \frac{1}{3} \times 3.142 \times 4675 \text{ cm}^2$
    $= 4896.283333 \text{ cm}^2$
    $= 4900 \text{ cm}^2$ to 3 significant figures

3. Let the angle be $\theta°$.
    Arc length $= \frac{\theta}{360} \times 2\pi r$ so $14 = \frac{\theta}{360} \times 2 \times 3.142 \times 10$
    Hence, $\theta = \frac{14 \times 360}{2 \times 3.142 \times 10} = 80.20369192$
    The angle subtended at the centre $= 80.2°$ to 1 decimal place.

4. a) Arc AE $= \frac{106}{360} \times 2 \times 3.142 \times 10$ m $= 18.50288889$ m
    Perimeter of the floor
    $= (18.50288889 + 21 + 16 + 21)$ m
    $= 76.50288889$ m
    $= 76.5$ m to 3 significant figures

    b) Area of sector AOE $= \frac{106}{360} \times 3.142 \times (10)^2 \text{m}^2$
       $= 92.514444444 \text{ m}^2$

    Area of trapezium ABCO
    $=$ area of trapezium EDCO $= \left(\frac{21+15}{2}\right) \times 8 \text{ m}^2$
    $= 144 \text{ m}^2$

    Area of the floor $= (92.51444444 + 144 + 144) \text{ m}^2$
    $= 380.5144444 \text{ m}^2$
    $= 381 \text{ m}^2$ to 3 significant figures

Check your progress 1

1. The lengths of the sides of the parallelogram are 6 cm, 5 cm, 6 cm and 5 cm. The perpendicular height is 4 cm.
    a) Perimeter $= (6 + 5 + 6 + 5)$ cm $= 22$ cm
    b) Area $=$ base $\times$ perpendicular height
       $= 6$ cm $\times 4$ cm
       $= 24 \text{ cm}^2$

2. a)

Rectangle	Area (cm²)	Perimeter (cm)
1st	$1 \times 2 = 2$	$2(1 + 2) = 6$
2nd	$3 \times 4 = 12$	$2(3 + 4) = 14$
3rd	$5 \times 6 = 30$	$2(5 + 6) = 22$
4th	$7 \times 8 = 56$	$2(7 + 8) = 30$

    b) Areas of next two rectangles $= (9 \times 10) \text{ cm}^2$ and $(11 \times 12) \text{ cm}^2$
       $= 90 \text{ cm}^2$ and 132 cm^2
       Perimeters of next two rectangles $= 2(9 + 10)$ cm and $2(11 + 12)$ cm
       $= 38$ cm and 46 cm

    c) Area of fifteenth rectangle $= (29 \times 30) \text{ cm}^2 = 870 \text{ cm}^2$
       Perimeter of fifteenth rectangle $= 2(29 \times 30)$ cm $= 118$ cm

3. Distance moved in one revolution $= 2 \times 3.142 \times 0.3$ m $= 1.8852$ m
    Distance car moves before tyres need replacing $= 1.8852 \times 7.5 \times 10^6$ m
    $= 14.139 \times 10^6$ m
    $= 1.41 \times 10^7$ m to 3 significant figures

4. a) Circumference $= \pi \times$ diameter so 12 m $= 3.142 \times$ diameter
       Diameter $= \frac{12}{3.142}$ m $= 3.819223425$ m
       Diameter $= 3.82$ m to 3 significant figures

    b) Area of cross-section $= \pi \times (1.2)^2 \text{ cm}^2 = 4.52448 \text{ cm}^2$
       $= 4.52 \text{ cm}^2$ to 3 significant figures

Module 5 Solutions

Check your progress 1 (cont.)

5. a) (i) Circumference = $\pi \times$ diameter = $2\pi x$ m
   (ii) Perimeter of track = $(2\pi x + 4x + 4x)$ m = $(2\pi x + 8x)$ m
   (iii) $(2\pi x + 8x) = 2x(\pi + 4)$
   b) (i) Perimeter = $2x(\pi + 4)$ m = 400 m
       Divide both sides by 2: $x(\pi + 4) = 200$
   (ii) Putting $\pi = 3.142$, $x(7.142) = 200$

   $$x = \frac{200}{7.142} = 28.0033604$$

   $x = 28.0$ to 3 significant figures
   (iii) AB = 112 m, BQC = 88 m, CD = 112 m, DPA = 88 m
   (Check: AB + BQC + CD + DPA = 400 metres.)

6. a) Let angle AOB = $\theta°$.   Area of sector = $\frac{\theta}{360} \times \pi r^2$

   $$60\pi = \frac{\theta}{360} \times \pi(12)^2$$

   $$\theta = \frac{60\pi \times 360}{\pi \times 144} = 150$$

   angle AOB = $150°$

   b) Length of arc = $\frac{\theta}{360} \times 2\pi r = \frac{150}{360} \times 2 \times \pi \times 12 = 10\pi$ cm

   Perimeter of sector = $(10\pi + 12 + 12)$ cm = $(10\pi + 24)$ cm

EXERCISE 10

1. a) Surface area = (56×45) mm$^2 \times 2 + (45 \times 36)$ mm$^2 \times 2 + (36 \times 56)$ mm$^2 \times 2$
       = $(5040 + 3240 + 4032)$ mm^2 = 12312 mm^2
   b) Surface area = (120×75) cm$^2 \times 2 + (75 \times 60)$ cm$^2 \times 2 + (60 \times 120)$ cm$^2 \times 2$
       = $(18\,000 + 9000 + 14\,400)$ cm^2 = 41 400 cm^2

2. Surface area = (31×16) cm$^2 \times 2 + (16 \times 5)$ cm$^2 \times 2 + (5 \times 31)$ cm$^2 \times 2$
       = $(992 + 160 + 310)$ cm^2 = 1462 cm^2

3. Depth of water = $\frac{3}{4} \times 20$ cm = 15 cm

   Area in contact with water = (30×20) cm$^2 + (20 \times 15)$ cm$^2 \times 2 + (30 \times 15)$ cm$^2 \times 2$
       = $(600 + 600 + 900)$ cm^2 = 2100 cm^2

4. a) A cube
   b) Area of each face = 1.5 cm $\times$ 1.5 cm = 2.25 cm^2
      Surface area of cube = 2.25 cm$^2 \times 6$ = 13.5 cm^2

EXERCISE 11

1. Curved surface area = $2\pi rh = 2 \times 3.142 \times 25 \times 80$ cm^2
       = 12 568 cm^2
   Area of the two ends = $\pi r^2 + \pi r^2$
       = $2 \times 3.142 \times (25)^2$ cm^2
       = 3927.5 cm^2
   Total surface area of tank = 16 495.5 cm^2
       = 16 500 cm^2 to 3 significant figures

2. Curved surface area = $2\pi rh = 2 \times 3.142 \times 11 \times 3$ mm^2
       = 207.372 mm^2
   Area of the plane faces = $\pi r^2 + \pi r^2$
       = $2 \times 3.142 \times (11^2)$ mm^2
       = 760.364 mm^2
   Total surface area of the coin
       = 967.736 mm^2
       = 968 mm^2 to 3 significant figures

3. Length of the label = $(2\pi r + 0.7)$ cm
       = $(2 \times 3.142 \times 3.7 + 0.7)$ cm
       = 23.9508 cm
   Area of the label = 23.9508×10.6 cm^2
       = 253.87848 cm^2
       = 254 cm^2 to 3 significant figures

EXERCISE 12

1. a) Volume = $(3 \times 2 \times 2)$ cm^3 = 12 cm^3
    b) Volume = $(3 \times 4 \times 3)$ cm^3 = 36 cm^3
    c) The base layer consists of 10 cubes and there are 3 layers.
       Volume = (10×3) cm^3 = 30 cm^3

2. a) Volume = 8.25 cm $\times$ 7.5 cm $\times$ 16 cm = 990 cm^3
    b) Volume = $4\frac{1}{2}$ cm $\times$ $3\frac{1}{3}$ cm $\times$ $1\frac{2}{5}$ cm
       = $\frac{9}{2} \times \frac{10}{3} \times \frac{7}{5}$ cm^3 = 21 cm^3

3. Volume = 80 cm $\times$ 75 cm $\times$ 0.3 cm = 1800 cm^3

4. a) Original volume = 350 mm $\times$ 90 mm $\times$ 80 mm
       = 2 520 000 mm^3 (or 2520 cm^3)
    b) Reduced volume = 300 mm $\times$ 85 mm $\times$ 70 mm
       = 1 785 000 mm^3
       Volume removed = $(2\,520\,000 - 1\,785\,000)$ mm^3
       = 735 000 mm^3 (or 735 cm^3)

EXERCISE 13

1. a) Volume = cross-sectional area $\times$ length
       = 1.4 cm^2 $\times$ 11 cm = 15.4 cm^3
    b) Volume = $\pi r^2 h$ = 3.142 $\times$ $(11)^2 \times 3$ mm^3
       = 1140.546 mm^3
       = 1140 mm^3 to 3 significant figures

2. a) Surface area = $(\frac{1}{2}(3 \times 4) + \frac{1}{2}(3 \times 4) + (4 \times 16)$ +
       $(5 \times 16) + (3 \times 16))$ cm^2
       = $(6 + 6 + 64 + 80 + 48)$ cm^3
       = 204 cm^2
    b) Volume = area of cross-section $\times$ length
       = 6 cm^2 $\times$ 16 cm = 96 cm^3

3. a) (i) Radius of well = 0.8 m
       (ii) Volume of well = $\pi r^2 h$ = 3.142 $\times$ $(0.8)^2 \times 10$ m^3
       = 20.1088 m^3
       = 20.1 m^3 to 3 significant figures
    b) Mass of rock removed = 20.1088 $\times$ 2.3 tonnes
       = 46.25024 tonnes
       = 46.3 tonnes to 3 significant figures

    > if you use the π key on your calculator, you obtain 46.2 tonnes as the answer to part b)

4. a) Area of cross-section = 6 cm^2
    b) Volume = area cross-section $\times$ length
       = 6 cm^2 $\times$ 300 cm = 1800 cm^3

5. Volume of one knitting needle
    = $\pi r^2 h$
    = 3.142 $\times$ $(0.3)^2 \times 32$ cm^3
    = 9.0489 cm^3
    Volume of plastic in 800 needles = 9.04896 $\times$ 800 cm^3
    = 7239.168 cm^3
    = 7240 cm^3 to 3 significant figures

EXERCISE 14

1. 0.3 ℓ = 300 mℓ
    Number of 5 mℓ spoonfuls = $\frac{300}{5}$ = 60

2. Volume of wine in 1 bottle = 75 cℓ = 750 mℓ
    So volume of wine in 8 bottles = 8 $\times$ 750 mℓ = 6000 mℓ
    $\frac{6000}{70}$ = 85.7
    So number of glasses that can be completely filled from 8 bottles is 85.

3. Capacity = $\pi r^2 h$ = 3.142 $\times$ $4^2 \times 12$ cm^3 = 603.264 cm^3
    = 603 cm^3 to 3 significant figures
    = 0.603 ℓ

Module 5 Solutions

EXERCISE 15

1. a) Volume = 150 cm × 100 cm × 0.2 cm = 3000 cm^3
   b) Mass = 3000 × 2.8 g = 8400 g

2. Area of cross-section = $(\pi \times 8^2 - \pi \times 6^2)$ mm^2
   = $(64 - 36)\pi$ mm^2
   = 28π mm^2
   Volume of the tube = (28π) mm^2 × 3000 mm
   = 84 000π mm^3 = 263 928 mm^3
   = 264 000 mm^3 (or 264 cm^3) to 3 significant figures

3. Total area of the ends = $(\pi \times 5^2 + \pi \times 5^2)$ cm^2
   = 50π cm^2
   Curved surface area = $(270 - 50\pi)$ cm^2 = 112.9 cm^2
   Curved surface area = $2\pi rh$ so $h = \frac{112.9}{2 \times 3.142 \times 5}$
   = 3.593252705
   Height of cylinder = 3.6 cm to the nearest millimetre.

4. a) Volume delivered in 1 minute = volume of cylinder with radius 0.8 cm and length (30 × 60) cm
   = $\pi \times (0.8)^2 \times 1800$ cm^3 = 3619.584 cm^3
   = 3620 cm^3 to 3 significant figures

   b) Area of the base of the tank = 25 cm × 20 cm
   = 500 cm^2
   Rise in the water level = $\frac{3619.584}{500}$ cm
   = 7.239168 cm
   = 7.2 cm to 2 significant figures

5. Volume of apple juice in the bottle = $\pi \times 6^2 \times 12$ cm^3
   = 432π cm^3
   Volume of apple juice in one beaker = $\pi \times 3^2 \times 7$ cm^3
   = 63π cm^3
   $\frac{\text{Volume in bottle}}{\text{Volume in beaker}} = \frac{432\pi}{63\pi} = 6.857142857$
   Number of beakers that can be filled = 6

EXERCISE 16

1. a) Volume of a sphere of ice-cream
   = $\frac{4}{3} \times 3.14 \times (2.5)^3$ cm^3
   = 65.41666667 cm^3
   = 65.4 cm^3 to 1 decimal place
   b) 1 ℓ = 1000 cm^3 and $\frac{1000}{65.14666667} = 15.286624$
   15 spheres of ice-cream can be made.

2. a) Triangular pyramid
   b) EF = 13 cm (the same length as AE)
   c) Volume = $\frac{1}{3}$ area of base × perpendicular height
   = $\frac{1}{3} \times (\frac{1}{2} \times 5 \times 5) \times 12$ cm^3 = 50 cm^3

3. a) Volume = $\frac{1}{3}\pi r^2 h$
   = $\frac{1}{3} \times 3.142 \times (1.2)^2 \times (1.5)$ m^3
   = 2.26224 m^3
   = 2.26 m^3 to 3 significant figures
   b) Curved surface area = $\pi r l$
   = $3.142 \times 1.2 \times 1.92$ m^2
   = 7.239168 m^2
   = 7.24 m^2 to 3 significant figures

4. a) Volume of the sphere = $\frac{4}{3}\pi r^3 = \frac{4}{3} \times 3.142 \times 5^3$ cm^3
   = $\frac{500\pi}{3}$ cm^3
   = 523.6666667 cm^3
   = 524 cm^3 to 3 significant figures
   b) (i) Radius of the box = 5 cm and height of the box = 10 cm.
   Total surface area of the box
   = $2\pi rh + \pi r^2 + \pi r^2$
   = $(2 \times \pi \times 5 \times 10) + (2 \times \pi \times 5^2)$ cm^2
   = 150π cm^2 = 471.3 cm^2
   = 471 cm^2 to 3 significant figures
   (ii) Capacity of the box = $\pi r^2 h = \pi \times 5^2 \times 10$ cm^3
   = 250π cm^3
   $\frac{\text{Volume of sphere}}{\text{Capacity of box}} = \frac{500\pi}{3} \div 250\pi = \frac{2}{3}$

EXERCISE 17

1. Linear scale factor $= \frac{16}{10} = \frac{8}{5}$
   Volume scale factor $= \left(\frac{8}{5}\right)^3 = \frac{512}{125}$
   Capacity of the larger jug $= 500 \text{ m}\ell \times \frac{512}{125} = 2048 \text{ m}\ell$

2. a) Volume of cone $= \frac{1}{3}\pi r^2 h$
   $= \frac{1}{3} \times 3.142 \times 6^2 \times 10 \text{ cm}^3 = 377.04 \text{ cm}^3$
   $= 377 \text{ cm}^3$ to 3 significant figures
   b) (i) Linear scale factor $= \frac{5}{10} = \frac{1}{2}$
   Radius of smaller piece $= \frac{6 \text{ cm}}{2} = 3 \text{ cm}$
   (ii) Volume of smaller piece $= \frac{1}{3} \times 3.142 \times 3^2 \times 5 \text{ cm}^3 = 47.13 \text{ cm}^3$
   $= 47.1 \text{ cm}^3$ to 3 significant figures
   Alternative method Volume scale factor = (linear scale factor)$^3 = \frac{1}{8}$
   Volume of smaller piece $= 377.04 \text{ cm}^3 \times \frac{1}{8} = 47.13 \text{ cm}^3$
   $= 47.1 \text{ cm}^3$ to 3 significant figures

3. a) (i) Surface area scale factor $= \frac{450}{200} = 2.25$
   (ii) Linear scale factor $= \sqrt{2.25} = 1.5$
   (iii) Volume scale factor = (linear scale factor)$^3 = (1.5)^3 = 3.375$
   b) Volume of the smaller solid $= \frac{1350 \text{ cm}^3}{3.375} = 400 \text{ cm}^3$

4. a) Area of cross-section $= ((9 \times 8) + \frac{1}{2}\pi(4)^2) \text{ cm}^2 = (72 + (8 \times 3.142)) \text{ cm}^2$
   $= 97.136 \text{ cm}^3$
   Volume of the loaf $= 97.136 \text{ cm}^2 \times 15 \text{ cm} = 1457.04 \text{ cm}^3$
   $= 1460 \text{ cm}^3$ to 3 significant figures
   b) (i) Volume scale factor $= (50)^3 = 125\,000$
   Volume of the balloon $= 1457.04 \text{ cm}^3 \times 12\,500 = 1.8213 \times 10^8 \text{ cm}^3$
   $= 1.82 \times 10^8 \text{ cm}^3$ (or 182 m^3) to 3 significant figures
   (ii) Area of each end of the loaf = area of cross-section $= 97.136 \text{ cm}^2$
   Perimeter of the cross-section $= (9 + 8 + 9 + (3.142 \times 4)) \text{ cm}$
   $= 38.568 \text{ cm}$
   Area of base and side of loaf $= 38.568 \text{ cm} \times 15 \text{ cm} = 578.52 \text{ cm}^2$
   Total surface area of the loaf $= (578.52 + 97.136 + 97.136) \text{ cm}^2$
   $= 772.792 \text{ cm}^2$
   Area scale factor of the enlargement $= (50)^2 = 2500$
   Surface area of the balloon $= 772.792 \text{ cm}^2 \times 2500 = 1\,931\,980 \text{ cm}^2$
   $= 1\,930\,000 \text{ cm}^2$ to 3 significant figures
   Area of material to make the balloon $= 193 \text{ m}^2$ to 3 significant figures.

Check your progress 2

1. a) $1 \ell = 1000 \text{ m}\ell$
   $\frac{1000}{120} = 8.333333333$ so number of full glasses $= 8$
   b) $120 \text{ m}\ell \times 8 = 960 \text{ m}\ell$
   Amount left over $= (1000 - 960) \text{ m}\ell = 40 \text{ m}\ell$

2. a) Length = diameter $\times 3 = 6.4 \text{ cm} \times 3 = 19.2 \text{ cm}$
   Breadth = diameter $\times 2 = 12.8 \text{ cm}$ and
   Height = diameter $= 6.4 \text{ cm}$
   b) Volume of box $= (19.2 \text{ cm} \times 12.8 \text{ cm} \times 6.4 \text{ cm})$
   $= 1572.864 \text{ cm}^3$
   $= 1570 \text{ cm}^3$ to 3 significant figures
   c) Surface area $= (19.2 \times 12.8) \text{ cm}^2 \times 2 +$
   $(12.8 \times 6.4) \text{ cm}^2 \times 2 +$
   $(6.4 \times 19.2) \text{ cm}^2 \times 2$
   $= (491.52 + 163.84 + 245.76) \text{ cm}^2$
   $= 901.12 \text{ cm}^2$
   $= 901 \text{ cm}^2$ to 3 significant figures

3. a) Radius $= 30 \text{ cm}$
   b) Volume $= \pi r^2 h = 3.142 \times (30)^2 \times 90 \text{ cm}^3$
   $= 254\,502 \text{ cm}^3$
   $= 255\,000 \text{ cm}^3$ to 3 significant figures
   c) (i) Curved surface area $= 2\pi rh$
   $= 2 \times 3.142 \times 30 \times 90 \text{ cm}^2$
   $= 16\,966.8 \text{ cm}^2$
   $= 17\,000 \text{ cm}^2$ to 3 significant figures
   (ii) Area of the base $= \pi r^2 = 3.142 \times (30)^2 \text{ cm}^2$
   $= 2827.8 \text{ cm}^2$
   $= 2830 \text{ cm}^2$ to 3 significant figures
   (iii) Total area of metal used
   $= (916\,966.8 + 2827.8) \text{ cm}^2$
   $= 19\,794.6 \text{ cm}^2$
   $= 1.97946 \text{ m}^2$ $(1 \text{ m}^2 = 10\,000 \text{ cm}^2)$
   $= 2.0 \text{ m}^2$ to 2 significant figures

Check your progress 2 (cont.)

d) (i) Area of rectangle ABCD $= 2.5 \text{ m} \times 0.9 \text{ m}$
$= 2.25 \text{ m}^2$
(ii) Area of metal wasted $= (2.25 - 2.0) \text{ m}^2$
$= 0.25 \text{ m}^2$
Percentage wasted $= \frac{0.25}{2.25} \times 100\%$
$= 11\%$ to 2 significant figures

4. a) Volume of prism = area of cross-section × length
$= \frac{1}{2} \times (90 \times 16) \times 300 \text{ cm}^3$
$= 216\,000 \text{ cm}^3$
b) Volume of cuboid $= (60 \times 16 \times 300) \text{ cm}^3$
$= 288\,000 \text{ cm}^3$
c) Volume of ramp $= (216\,000 + 288\,000) \text{ cm}^3$
$= 504\,000 \text{ cm}^3$

5. The two hemispherical ends together can be regarded as a sphere as far as curved surface area and volume are concerned.
a) Total surface area $= 4\pi r^2 + 2\pi r h$
$= ((4 \times \pi \times 9^2) + (2 \times \pi \times 9 \times 100)) \text{ cm}^2$
$= 2124\pi \text{ cm}^2 = 6673.608 \text{ cm}^2$
$= 6670 \text{ cm}^2$ to 3 significant figures
b) Total volume $= \frac{4}{3}\pi r^3 + \pi r^2 h$
$= ((\frac{4}{3} \times \pi \times 9^3) + (\pi \times 9^2 \times 100))$
$= 9072\pi \text{ cm}^3 = 28\,504.224 \text{ cm}^3$
$= 28\,500 \text{ cm}^3$ to 3 significant figures

EXERCISE 18

1. a) $a^2 = (15)^2 + (36)^2 = 225 + 1296 = 1521$
$a = \sqrt{1521} \text{ cm} = 39 \text{ cm}$
b) $(25)^2 = b^2 + 7^2$ so $b^2 = (25)^2 - 7^2$
$= 625 - 49 = 576$
$b = \sqrt{576} \text{ cm} = 24 \text{ cm}$
c) $c^2 = 3^2 + (1.6)^2 = 9 + 2.56 = 11.56$
$c = \sqrt{11.56} \text{ m} = 3.4 \text{ m}$
d) $(65)^2 = d^2 + (60)^2$ so $d^2 = (65)^2 - (60)^2$
$= 4225 - 3600 = 625$
$d = \sqrt{625} \text{ mm} = 25 \text{ mm}$

2. $x^2 + 7^2 = 9^2$ so $x^2 = 9^2 - 7^2 = 81 - 49 = 32$
$x = \sqrt{32} \text{ cm} = 5.65685425 \text{ cm}$
$x = 5.66 \text{ cm}$ to 3 significant figures

3. $(BC)^2 + (7.5)^2 = (8.5)^2$ so $(BC)^2 = (8.5)^2 - (7.5)^2$
$= 72.25 - 56.25 = 16$
$BC = \sqrt{16} \text{ km} = 4 \text{ km}$

4.

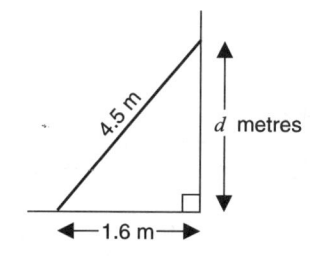

Let the distance be d metres
$d^2 + (1.6)^2 = (4.5)^2$
$d^2 = (4.5)^2 - (1.6)^2$
$= 20.25 - 2.56 = 17.69$
$d = \sqrt{17.69} = 4.205948169$

Distance ladder reaches up the wall
$= 4.21 \text{ m}$ to 3 significant figures.

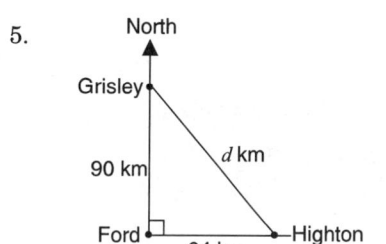

5.

Let the distance be d kilometres
$d^2 = (90)^2 + (64)^2$
$= 8100 + 4096$
$= 12\,196$
$d = \sqrt{12\,196} = 110.4355015$

Distance from Highton to Grisley
$= 110 \text{ km}$ to 3 significant figures.

6. a) Angle $OTP = 90°$ because it is the angle between a tangent and the radius through the point of contact.
b) $(PT)^2 + (OT)^2 = (OP)^2$
so $(PT)^2 = (OP)^2 - (OT)^2 = 9^2 - 4^2 = 81 - 16$
$(PT)^2 = 65$
$PT = \sqrt{65} \text{ cm} = 8.062257748 \text{ cm}$
$PT = 8.06 \text{ cm}$ to 3 significant figures

7. a) $(AP)^2 = (3.2)^2 + (2.0)^2 = 10.24 + 4.00 = 14.24$
$AP = \sqrt{14.24} \text{ m} = 3.773592453 \text{ m}$
$AP = 3.77 \text{ m}$ to 3 significant figures
b) In triangle BPC, angle $C = 90°$,
$PC = (5.8 - 2.0) \text{ m} = 3.8 \text{ m}, BC = 3.2 \text{ m}.$
$(BP)^2 = (3.8)^2 + (3.2)^2 = 14.44 + 10.24 = 24.68$
$BP = \sqrt{24.68} \text{ m} = 4.967896939 \text{ m}$
so $BP = 4.97 \text{ m}$ to 3 significant figures

EXERCISE 19

1. a) The diagram shows the equilateral triangle ABC.
   AN is the perpendicular from A to BC.
   BN = CN = 5 cm.

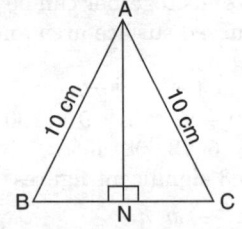

   Using Pythagoras's theorem in triangle ABN
   $(AB)^2 = (AN)^2 + (BN)^2$
   $(10)^2 = (AN)^2 + (5)^2$
   $(AN)^2 = (10)^2 - (5)^2 = 100 - 25$
   $AN = \sqrt{75}$ cm = 8.660254038 cm
   Perpendicular height of triangle
   = 8.66 cm to 3 significant figures.

   b) Area of triangle ABC
   = $\frac{1}{2}$ × base × perpendicular height
   = $\frac{1}{2}$ × 10 cm × 8.660254038 cm
   = 43.30127019 cm^2
   = 43.3 cm^2 to 3 significant figures

2. a) AC is a diagonal of the face ABCD. Triangle ABC has a right angle at C.
   $(AC)^2 = (20)^2 + (20)^2$
   = 400 + 400
   = 800
   $AC = \sqrt{800}$ = 28.28427125
   Diagonal of a face = 28.3 cm to 3 significant figures.

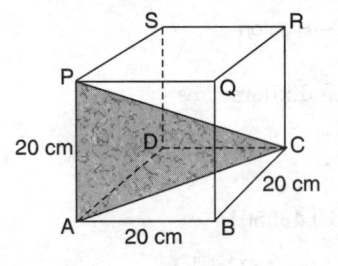

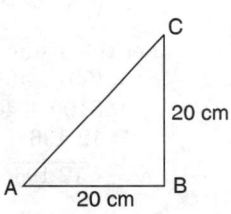

   b) PC is a diagonal of the cube.
   Triangle PAC has a right angle at A.
   $(PC)^2 = (AP)^2 + (AC)^2$
   = 400 + 800 = 1200
   $PC = \sqrt{1200}$ cm = 34.64101615 cm
   Diagonal of the cube = 34.6 cm to 3 significant figures.

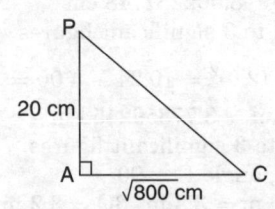

3. CN is the perpendicular from C to AB.
   $AN = NB = \frac{1}{2}(48 \text{ cm}) = 24$ cm
   $(AC)^2 = (CN)^2 + (AN)^2$
   $(25)^2 = (CN)^2 + (24)^2$
   $(CN)^2 = (25)^2 - (24)^2 = 625 - 576 = 49$
   $CN = \sqrt{49}$ cm = 7 cm
   Perpendicular distance from C to AB = 7 cm.

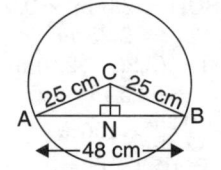

4. a) Triangle EAB has a right angle at A.
   $(EB)^2 = (EA)^2 + (AB)^2 = 3^2 + 4^2 = 9 + 16 = 25$
   $EB = \sqrt{25}$ cm
   EB = 5 cm

   b) Triangle ABC has a right angle at C.
   $(AC)^2 = (AB)^2 + (BC)^2 = 4^2 + 4^2 = 16 + 16 = 32$
   $AC = \sqrt{32}$ cm = 5.65685425 cm
   AC = 5.66 cm to 3 significant figures

   c) Triangle EAC is right-angled at A. (EA is vertical and AC is horizontal.)
   $(EC)^2 = (EA)^2 + (AC)^2 = 9 + 32 = 41$
   $EC = \sqrt{41}$ cm = 6.403124237 cm
   EC = 6.40 cm to 3 significant figures

5. a) (Longest side)$^2 = (13.6)^2 = 184.96$
   Sum of squares of the other two sides
   $= (6.4)^2 + (12)^2$
   = 40.96 + 144 = 184.96
   $(13.6)^2 = (6.4)^2 + (12)^2$ so the triangle is right-angled.

   b) (Longest side)$^2 = (110)^2 = 12\,100$
   Sum of the squares of the other two sides
   $= (85)^2 + (65)^2$
   = 7225 + 4225 = 11 450
   $(110)^2 \neq (85)^2 + (65)^2$ so the triangle is not right-angled.

   c) (Longest side)$^2 = 5^2 = 25$
   Sum of the squares of the other two sides
   $= (2.7)^2 + (4.2)^2$
   = 7.29 + 17.64 = 24.93
   $5^2 \neq (2.7)^2 + (4.2)^2$ so the triangle is not right-angled.

   d) (Longest side)$^2 = (10)^2 = 100$
   Sum of the squares of the other two sides
   $= (2.8)^2 + (9.6)^2$
   = 7.84 + 92.16 = 100
   $(10)^2 = (2.8)^2 + (9.6)^2$ so the triangle is right-angled.

6. (Longest side)$^2 = (82)^2 = 6724$
   Sum of the squares on the other two sides
   $= (53)^2 + (65)^2$
   = 2809 + 4225 = 7034
   $(82)^2$ is less than $(53)^2 + (65)^2$, and hence the greatest angle of the triangle is less than 90°.
   The triangle is acute-angled.

Module 5 Solutions

EXERCISE 20

1. a) hypotenuse = 20   opp(A) = 16   adj(A) = 12
   b) hypotenuse = 25   opp(A) = 7   adj(A) = 24
   c) hypotenuse = 13   opp(A) = 12   adj(A) = 5
   d) hypotenuse = 85   opp(A) = 84   adj(A) = 13

2. PR is the hypotenuse, RQ is opp(50°), PQ is adj(50°).

3. hypotenuse = 200   opp(60°) = 173   adj(60°) = 100
                      opp(30°) = 100   adj(30°) = 173

EXERCISE 21

1. a) We assume you have drawn the diagram on graph paper.
   b) $B_1C_1 = 2.3$ to 2.4 cm $B_2C_2 = 4.7$ to 4.8 cm
      $B_3C_3 = 7.1$ to 7.2 cm $B_4C_4 = 9.5$ to 9.6 cm
      $B_5C_5 = 11.9$ to 12.0 cm
   c) Each of the values should be 1.2 to 2 significant figures.
   d) Estimated value of $\tan(50°) = 1.2$.

2. a) Display is

   | DEG | 0.267949192 |

   so $\tan(15°) = 0.2679$ to 4 decimal places.

   b) Display is

   | DEG | 0.767326988 |

   so $\tan(37.5°) = 0.7673$ to 4 decimal places.

   c) Display is

   | DEG | 1.303225373 |

   so $\tan(52.5°) = 1.3032$ to 4 decimal places.

   d) Display is

   | DEG | 5.144554016 |

   so $\tan(79°) = 5.1446$ to 4 decimal places.

3. a) Display is

   | DEG | 0.017455064 |

   so $\tan(1°) = 0.01746$ to 4 significant figures.

   b) Display is

   | DEG | 0.17632698 |

   so $\tan(10°) = 0.1763$ to 4 significant figures.

   c) Display is

   | DEG | 1.053780125 |

   so $\tan(46.5°) = 1.054$ to 4 significant figures.

   d) Display is

   | DEG | 10.01870799 |

   so $\tan(84.3°) = 10.02$ to 4 significant figures.

4. a) $\tan(P) = \frac{\text{opp}(P)}{\text{adj}(P)} = \frac{8}{15}$
   b) $\tan(P) = 0.533333333$
      $= 0.5333$ to 4 decimal places

5. a) $\tan(65°) = \frac{\text{opp}(65°)}{\text{adj}(65°)}$
      $= \frac{107}{49.9} = 2.144288577$
      $= 2.14$ to 3 significant figures
   b) $\tan(25°) = \frac{\text{opp}(25°)}{\text{adj}(25°)}$
      $= \frac{49.9}{107} = 0.46635514$
      $= 0.466$ to 3 significant figures

6. a) $\tan(47°) = 1.07236871$
      $= 1.0724$ to 4 decimal places
   b) $\tan(47°) = \frac{\text{opp}(47°)}{\text{adj}(47°)} = \frac{OT}{30}$
      $OT = 30 \times \tan(47°)$
      $= 32.1710613$
   Height of the tree $= 32.2$ m to 3 significant figures.

7. $\tan(22°) = \frac{\text{opp}(22°)}{\text{adj}(22°)} = \frac{TA}{80}$
   $TA = 80 \times \tan(22°) = 32.32209807$
   Width of the river $= 32.3$ m to 3 significant figures.

8. a) If A is acute and less than 45°, tan(A) is between 0 and 1.
   b) If A is acute and greater than 45°, tan(A) is greater than 1.

9. a) (i) $BD = 2$ units and the triangle is equilateral, so $AB = 2$ units.
      (ii) $(AC)^2 + (BC)^2 = (AB)^2$
           $(AC)^2 = (AB)^2 - (BC)^2$
           $= 2^2 - 1^2 = 4 - 1 = 3$
           $AC = \sqrt{3}$
   b) exact value of $\frac{BC}{AC} = \frac{1}{\sqrt{3}}$
   c) $\tan(30°) = \frac{1}{\sqrt{3}}$
      $= \frac{1}{1.732050808}$
      $= 0.577350269$
      $= 0.5774$ to 4 significant figures

EXERCISE 22

1. a) $\tan^{-1}(0.85) = 40.36453657° = 40.4°$ to 1 decimal place
    b) $\tan^{-1}(1.2345) = 50.99098611° = 51.0°$ to 1 decimal place
    c) $\tan^{-1}(3.56) = 74.31000707° = 74.3°$ to 1 decimal place
    d) $\tan^{-1}(10) = 84.28940686° = 84.3°$ to 1 decimal place

2. a) Press the keys [2] [÷] [5] [=] [SHIFT] [tan⁻¹].
    The display is

    | DEG   21.80140949 |

    so the angle is 22° (nearest degree).

    b) Press the keys [7] [÷] [9] [=] [SHIFT] [tan⁻¹].
    The display is

    | DEG   37.87498365 |

    so the angle is 38° (nearest degree).

    c) Press the keys
    [2] [5] [÷] [3] [2] [=] [SHIFT] [tan⁻¹]
    The display is

    | DEG   37.99873244 |

    so the angle is 38° (nearest degree).

    d) Press the keys
    [3] [÷] [4] [=] [+] [2] [=] [SHIFT] [tan⁻¹].
    or the keys
    [1] [1] [÷] [4] [=] [SHIFT] [tan⁻¹]
    or the keys
    [2] [.] [7] [5] [SHIFT] [tan⁻¹].
    The display is

    | DEG   70.01689348 |

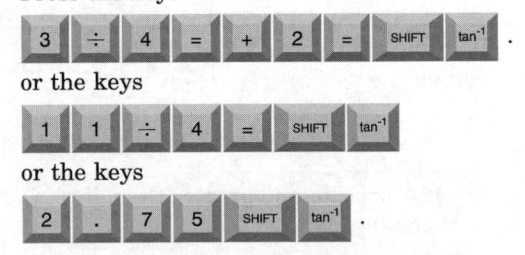

    so the angle is 70° (nearest degree).

3. a) $\tan(a) = \frac{7}{10}$ so $a = \tan^{-1}(0.7) = 34.9920202°$
    $= 35.0°$ to 1 decimal place
    b) $\tan(b) = \frac{9}{2}$ so $b = \tan^{-1}(4.5) = 77.47119229°$
    $= 77.5°$ to 1 decimal place
    c) $\tan(c) = \frac{4}{5}$ so $c = \tan^{-1}(0.8) = 38.65980825°$
    $= 38.7°$ to 1 decimal place
    $\tan(d) = \frac{5}{4}$ so $d = \tan^{-1}(1.25) = 51.34019175°$
    $= 51.3°$ to 1 decimal place
    Alternatively, $d = 180° - (90° + 38.7°)$
    $= 51.3°$ to 1 decimal place

4. If A is the angle the ladder makes with the ground,
    $\tan(A) = \frac{\text{opp}(A)}{\text{adj}(A)} = \frac{8.5}{2.8} = 3.035714286$ and so
    $A = 71.76750892°$
    The ladder makes an angle of 71.8° with the ground correct to 1 decimal place.

5. If A is the angle of elevation,
    $\tan(A) = \frac{\text{opp}(A)}{\text{adj}(A)} = \frac{68}{175} = 0.388571428$ and so
    $A = 21.23470413°$
    The angle of elevation $= 21.2°$ correct to 1 decimal place.

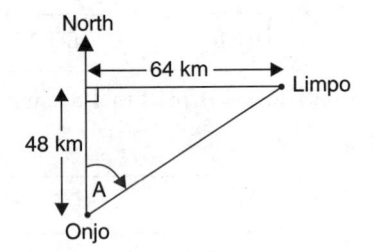

6.
    $\tan(A) = \frac{64}{48} = 1.333333333$
    $A = \tan^{-1}(1.333333333)$
    $= 53.13010235°$
    The bearing of Limpo from Onjo $= 053°$.

EXERCISE 23

1. a) (i) $\sin(A) = \frac{\text{opp}(A)}{\text{hypotenuse}} = \frac{21}{29}$
       (ii) $\cos(A) = \frac{\text{adj}(A)}{\text{hypotenuse}} = \frac{20}{29}$
    b) (i) $\sin(A) = \frac{\text{opp}(A)}{\text{hypotenuse}} = \frac{8}{17}$
       (ii) $\cos(A) = \frac{\text{adj}(A)}{\text{hypotenuse}} = \frac{15}{17}$
    c) (i) $\sin(A) = \frac{\text{opp}(A)}{\text{hypotenuse}} = \frac{12}{15}$
       (ii) $\cos(A) = \frac{\text{adj}(A)}{\text{hypotenuse}} = \frac{9}{15}$
    d) (i) $\sin(A) = \frac{\text{opp}(A)}{\text{hypotenuse}} = \frac{13}{85}$
       (ii) $\cos(A) = \frac{\text{adj}(A)}{\text{hypotenuse}} = \frac{84}{85}$

2. a) $\sin(5°) = 0.087155742$
    $= 0.0872$ to 4 decimal places
    b) $\sin(30°) = 0.5$ exactly
    c) $\sin(60°) = 0.866025403$
    $= 0.8660$ to 4 decimal places
    d) $\sin(85°) = 0.996194698$
    $= 0.9962$ to 4 decimal places

EXERCISE 23 (cont.)

3. a) $\cos(5°) = 0.996194698$
   $= 0.9962$ to 4 decimal places
   b) $\cos(30°) = 0.866025403$
   $= 0.8660$ to 4 decimal places
   c) $\cos(60°) = 0.5$ exactly
   d) $\cos(85°) = 0.087155742$
   $= 0.0872$ to 4 decimal places

4. a) $\cos(48°) = \frac{q}{r}$ b) $\sin(30°) = \frac{e}{f}$
   c) $\cos(35°) = \frac{HI}{IJ}$ d) $\cos(\theta) = \frac{x}{r}$

5. a) $\cos(28°) = \frac{a}{12}$ so $a = 12 \times \cos(28°)$
   $= 10.59537111$
   $= 10.6$ to 3 significant figures
   b) $\sin(25°) = \frac{b}{10}$ so $b = 10 \times \sin(25°)$
   $= 4.22618261$
   $= 4.23$ to 3 significant figures
   c) $\sin(70°) = \frac{c}{15}$ so $c = 15 \times \cos(70°)$
   $= 14.09538931$
   $= 14.1$ to 3 significant figures
   d) $\cos(32°) = \frac{45}{d}$ so $d \times \cos(32°) = 45$
   and $d = \frac{45}{\cos(32°)}$
   Hence, $d = 53.06302815 = 53.1$ to 3 significant figures.

6. a) $\sin^{-1}(0.99) = 81.89038554°$
   $= 81.9°$ to 1 decimal place
   b) $\cos^{-1}(0.5432) = 57.09825644°$
   $= 57.1°$ to 1 decimal place
   c) $\sin^{-1}(\frac{3}{8}) = \sin^{-1}(0.375)$
   $= 22.02431284°$
   $= 22.0°$ to 1 decimal place
   d) $\cos^{-1}(\frac{10}{23}) = \cos^{-1}(0.434782608)$
   $= 64.22853826°$
   $= 64.2°$ to 1 decimal place

7. a) $\sin(A) = \frac{7}{16}$ so $A = \sin^{-1}(\frac{7}{16})$
   $= 25.94447977°$
   $= 26°$ to nearest degree
   b) $\sin(B) = \frac{12}{17}$ so $B = \sin^{-1}(\frac{12}{17})$
   $= 44.90087216°$
   $= 45°$ to nearest degree

   c) $\cos(C) = \frac{7}{20}$ so $C = \cos^{-1}(\frac{7}{20})$
   $= 69.51268489°$
   $= 70°$ to nearest degree
   d) Use either $\sin(D) = \frac{60}{61}$ or $\cos(D) = \frac{11}{61}$
   or $\tan(D) = \frac{60}{61}$.
   These give $D = 79.61114281°$ so $D = 80°$ to the nearest degree.

8. $\sin(18°) = \frac{BC}{6.25}$
   so $BC = 6.25 \times \sin(18°) = 1.931356215$
   Difference in height between A and B = 1.93 m to 3 significant figures.

9. a) $\cos(56°) = \frac{PR}{18}$
   $PR = 18 \times \cos(56°)$
   $= 10.06547226$
   Distance Q is north of P
   $= 10.1$ km to 3 significant figures.
   b) $\sin(56°) = \frac{RQ}{18}$
   $RQ = 18 \times \sin(56°)$
   $= 14.92267631$
   Distance Q is east of P
   $= 14.9$ km to 3 significant figures.

10. $\sin(x°) = \frac{\text{opp}(x°)}{\text{hypotenuse}} = \frac{2050 \text{ m}}{5000 \text{ m}} = 0.41$
    $x° = \sin^{-1}(0.41) = 24.2048348°$
    Angle of slope of the track = 24.2° to 1 decimal place.

EXERCISE 24

1. a) $\sin(A\hat{B}C) = \frac{\text{opposite}}{\text{hypotenuse}} = \frac{5.2}{18.6} = 0.279569892$
   $A\hat{B}C = \sin^{-1}(0.279569892) = 16.23453623°$
   angle ABC = 16.2° to 1 decimal place
   b) Using Pythagoras's theorem in triangle ABC:
   $(AB)^2 = (BC)^2 + (AC)^2$
   $(AB)^2 - (AC)^2 = (BC)^2$
   $(BC)^2 = (18.6)^2 - (5.2)^2$
   $= 345.96 - 27.04$
   $= 318.92$
   $BC = \sqrt{318.92} = 17.85833139$
   Length of BC = 17.9 m to 3 significant figures.

   Alternatively, $\cos(A\hat{B}C) = \frac{BC}{18.6}$
   so $BC = 18.6 \times \cos(16.23453623°)$ m
   $= 17.9$ m to 3 significant figures

2. Angle $AON = \frac{120°}{2} = 60°$
   In triangle AON, $\sin(60°) = \frac{AN}{8}$
   $AN = 8 \times \sin(60°) = 6.92820323$
   $AN = BN$ so $AB = 6.92820323 \times 2$
   $= 13.85640646$
   Length of AB = 13.9 cm to 3 significant figures.

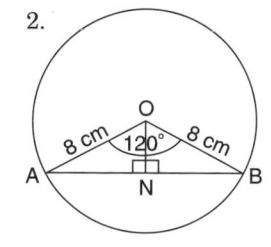

EXERCISE 24 (cont.)

3. a) In triangle ABC, angle C = 90°, AC = 1.5 m and BC = 0.9 m.
    $\tan(A\hat{B}C) = \frac{\text{opposite}}{\text{adjacent}} = \frac{1.5}{0.9} = 1.666666667$
    angle ABC = $\tan^{-1}(1.666666667) = 59.03624347°$
    Angle between AB and BD = 59.0° to 1 decimal place.
    b) Using Pythagoras's theorem in triangle ABC,
    $(AB)^2 = (AC)^2 + (BC)^2$
    $= (1.5)^2 + (0.9)^2$
    $= 2.25 + 0.81 = 3.06$
    $AB = \sqrt{3.06} = 1.749285568$
    Length of AB = 1.75 m to 3 significant figures.
    c) Volume inside the tent = area of cross-section × length
    $= (\tfrac{1}{2} \times 1.8 \times 1.5) \times 3 \text{ m}^3$
    Volume inside the tent = 4.05 m³

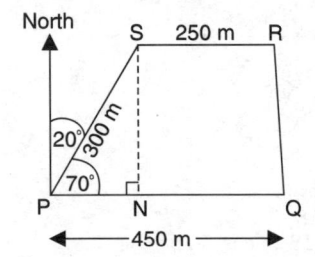

4.
    a) Bearing of S from P = 020°.
    b) In right-angled triangle SPN,
    $\sin(70°) = \frac{\text{opposite}}{\text{hypotenuse}} = \frac{SN}{300}$
    $SN = 300 \times \sin(70°)$
    $= 281.9077862$
    Shortest distance between SR and PQ
    = 282 m to 3 significant figures.

    c) PQ and SR are both due east so they are parallel, and PQRS is a trapezium. Area of PQRS
    = average of parallel sides × perpendicular distance between them
    $= \left(\frac{250 + 450}{2}\right) \times 281.9077862 \text{ m}^2$
    $= 98667.72518 \text{ m}^2$
    $= 98700 \text{ m}^2$ to 3 significant figures

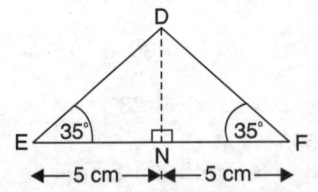

5.
    DN is the perpendicular from D to EF.
    $EN = NF = \frac{10 \text{ cm}}{2} = 5 \text{ cm}$
    a) In right-angled triangle DEN,
    $\tan(35°) = \frac{\text{opposite}}{\text{adjacent}} = \frac{DN}{5}$
    $DN = 5 \times \tan(35°) = 3.501037691$
    Perpendicular distance from D to EF
    = 3.50 cm to 3 significant figures.
    b) In right-angled triangle DEN,
    $\cos(35°) = \frac{\text{adjacent}}{\text{hypotenuse}} = \frac{5}{DE}$
    $DE \times \cos(35°) = 5$
    $DE = \frac{5}{\cos(35°)} = 6.103872944$
    Length of DE = 6.10 cm to 3 significant figures.

    Alternatively, using Pythagoras's theorem in triangle DEN,
    $(DE)^2 = (5)^2 + (3.501037691)^2 = 37.25726491$
    $DE = \sqrt{37.25726491} = 6.103872944$
    Length of DE = 6.10 cm to 3 significant figures.

Check your progress 3

1. a) $15 \cos(40°) = 11.49066665$
    b) $15 \cos(40°) = 11.5$ to 3 significant figures

2. $\cos(35°) = \frac{\text{adj}(35°)}{\text{hypotenuse}} = \frac{AC}{12}$ so
    $AC = 12 \times \cos(35°) = 9.829824531$ m
    AC = 9.83 m to 3 significant figures
    $\sin(35°) = \frac{\text{opp}(35°)}{\text{hypotenuse}} = \frac{BC}{12}$ so
    $BC = 12 \times \sin(35°) = 6.882917236$ m
    BC = 6.88 m to 3 significant figures

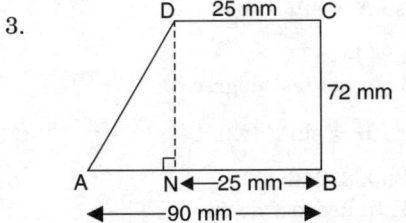

3.
    Draw DN, the perpendicular from D to AB. In right-angled triangle AND,
    AN = (90 − 25) mm = 65 mm and DN = 72 mm.
    Using Pythagoras's theorem,
    $(AD)^2 = (AN)^2 + (DN)^2 = (65)^2 + (72)^2$
    $= 4225 + 5184$
    $= 9409$
    $AD = \sqrt{9409}$ mm = 97 mm
    Perimeter of ABCD = (90 + 72 + 25 + 97) mm
    = 284 mm

Check your progress 3 (cont.)

4.

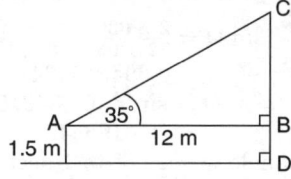

In right-angled triangle ABC,
$\tan(35°) = \frac{\text{opp}(35°)}{\text{adj}(35°)} = \frac{BC}{12}$
$BC = 12 \times \tan(35°) = 8.402490458$ m

Height of chimney
$= BC + $ height of girl's eyes
$= (8.402490458 + 1.5)$ m
$= 9.902490458$ m
$= 9.90$ m to 3 significant figures

5.

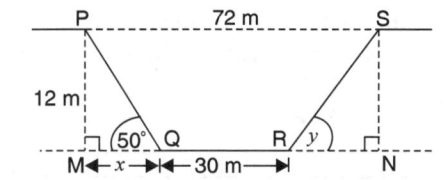

a) In right-angled triangle PQM,
angle $QPM = 180° - (90° + 50°) = 40°$
$\tan(Q\hat{P}M) = \tan(40°) = \frac{\text{opp}(40°)}{\text{adj}(40°)} = \frac{x}{12}$
$x = 12 \times \tan(40°)$ m
$= 10.06919557$ m
Horizontal distance between P and Q
$= 10.1$ m to 3 significant figures.

b) In right-angled triangle SRN,
$SN = 12$ m and $RN = (72 - (30 + x))$ m
$= 31.93080443$ m
$\tan(y) = \frac{SN}{RN} = \frac{12}{31.93080443} = 0.375812642$
$y = \tan^{-1}(0.375812642) = 20.59685492°$

Angle which RS makes with the horizontal
$= 20.6°$ to 1 decimal place.

6. a) (i) In right-angled triangle PQX,
$\tan(40°) = \frac{\text{opp}(40°)}{\text{adj}(40°)} = \frac{QX}{60}$ so
$QX = 60 \times \tan(40°)$ m $= 50.34597787$ m
Distance $QX = 50.3$ m to 3 significant figures.

ii) In right-angled triangle PQX,
$\cos(40°) = \frac{\text{adj}(40°)}{\text{hypotenuse}} = \frac{60}{PQ}$ so
$PQ = \frac{60}{\cos(40°)} = 78.32443736$ m
Distance of lion from the game warden
$= 78.3$ m to 3 significant figures.

Alternatively, using Pythagoras's theorem
$(PQ)^2 = (PX)^2 + (QX)^2$
$= (60)^2 + (50.34597787)^2$
$= 6134.717488$
$PQ = \sqrt{6134.717488}$ m
$= 78.32443736$ m
$= 78.3$ m to 3 significant figures

b) (i) $XR = XQ + QR = (50.34597787 + 200)$ m
$= 250.34597787$ m
$XR = 250.3$ m

(ii)

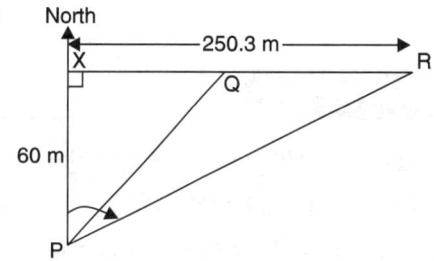

Using Pythagoras's theorem in triangle PRX
$(PR)^2 = (PX)^2 + (XR)^2$
$= (60)^2 + (250.34597787)^2$
$= 66\,273.10864$
$PR = \sqrt{66\,273.10864}$ m
$= 257.4356398$ m
Distance of the second lion from the game
warden $= 257$ m to 3 significant figures.

(iii) In right-angled triangle RPX, we can use
$\tan(P) = \frac{250.34597787}{60}$ or
$\sin(P) = \frac{250.34597787}{257.4356398}$ or
$\cos(P) = \frac{60}{257.4356398}$

Each of these gives $P = 76.522236\ldots°$
Bearing of second lion from the game warden
$= 077°$ (to nearest degree).

EXERCISE 25

1. a) $\sin 145° = 0.573576436$
$= 0.5736$ to 4 decimal places
b) $\cos 145° = -0.819152044$
$= -0.8192$ to 4 decimal places
c) $\sin 150° = 0.5$ exactly
d) $\cos 150° = -0.866025403$
$= -0.8660$ to 4 decimal places

2. a) Area $= \frac{1}{2}bc \sin A = \frac{1}{2} \times 5 \times 7 \times \sin 80°$
$= 17.23413568$ cm^2
$= 17.2$ cm^2 to 3 significant figures

b) Area $= \frac{1}{2}de \sin F = \frac{1}{2} \times 8.4 \times 5.5 \times \sin 100°$
$= 22.74905909$ cm^2
$= 22.7$ cm^2 to 3 significant figures

c) Area $= \frac{1}{2} \times 7 \times 8 \times \sin 120°$
$= 24.24871131$ cm^2
$= 24.2$ cm^2 to 3 significant figures

d) Area $= ((\frac{1}{2} \times 5 \times 15 \times \sin 127°) +$
$(\frac{1}{2} \times 13 \times 9 \times \sin 112.5°))$ cm^2
$= (29.94883163 + 54.04695265)$ cm^2
$= 83.99578428$ cm^2
$= 84.0$ cm^2 to 3 significant figures

EXERCISE 25 (cont.)

3. Area of triangle BCD = area of triangle ABD
$= \frac{1}{2} \times 9 \times 12 \times \sin 95° \, \text{cm}^2$
Area of parallelogram ABCD $= 9 \times 12 \times \sin 95° \, \text{cm}^2$
$= 107.5890274 \, \text{cm}^2$
$= 108 \, \text{cm}^2$ to 3 significant figures

4. a) (i) $A = \sin^{-1}(0.83) = 56.098738°$
   $= 56°$ to the nearest degree
   (ii) $A = 180° - 56° = 124°$ to the nearest degree
   b) $\cos^{-1}(-0.48) = 118.685402°$
   $= 118.7$ to 1 decimal place

5. a) Angle Q is opposite the shortest side so it must be the smallest angle in the triangle. Hence, angle Q is acute.

   b) Area $= \frac{1}{2}pr \sin Q$ so $630 = \frac{1}{2} \times 52 \times 63 \times \sin Q$
   $\sin Q = \frac{2 \times 630}{52 \times 63}$
   $= 0.384615384$
   $Q = \sin^{-1}(0.384615384)$
   $= 22.61986495°$
   angle $Q = 22.6°$ to 1 decimal place

   c) Side PQ is the longest side in the triangle so the opposite angle, R, is the largest angle. Angle P is not the largest angle and so it must be acute.
   Area of triangle $= \frac{1}{2}qr \sin P$ so
   $630 = \frac{1}{2} \times 25 \times 63 \times \sin P$
   $\sin P = \frac{2 \times 630}{25 \times 63}$
   $= 0.8$
   $P = \sin^{-1}(0.8) = 53.13010235°$
   angle $P = 53.1°$ to 1 decimal place

EXERCISE 26

1. Angle $C = 180° - (72° + 45°) = 63°$
   $\frac{a}{\sin A} = \frac{b}{\sin B} = \frac{c}{\sin C}$ becomes $\frac{BC}{\sin 72°} = \frac{AC}{\sin 45°} = \frac{20}{\sin 63°}$
   Hence, $AC = \frac{20 \times \sin 45°}{\sin 63°} = 15.87208987$ and
   $BC = \frac{20 \times \sin 72°}{\sin 63°} = 21.34791263$
   To 3 significant figures, $AC = 15.9$ cm and $BC = 21.3$ cm.

2. Angle $F = 180° - (140° + 15°) = 25°$
   $\frac{d}{\sin D} = \frac{e}{\sin E} = \frac{F}{\sin F}$ becomes
   $\frac{EF}{\sin 140°} = \frac{6}{\sin 15°} = \frac{DE}{\sin 25°}$
   Hence, $DE = \frac{6 \times \sin 25°}{\sin 15°} = 9.797229448$ and
   $EF = \frac{6 \times \sin 140°}{\sin 15°} = 14.90124367$
   To 3 significant figures, $DE = 9.80$ m and $EF = 14.9$ m.

3. $\frac{\sin P}{p} = \frac{\sin Q}{q} = \frac{\sin R}{r}$ becomes $\frac{\sin P}{QR} = \frac{\sin 120°}{13} = \frac{\sin R}{8}$
   hence, $\sin R = \frac{8 \times \sin 120°}{13} = 0.53293871$
   angle P + angle R $= 180° - 120° = 60°$
   so P and R are both acute angles.
   Hence, $R = \sin^{-1}(0.53293871) = 32.2042275°$
   angle $R = 32.2°$ to 1 decimal place
   angle $P = 60° -$ angle $R = 27.7957725°$
   angle $P = 27.8°$ to 1 decimal place
   $\frac{p}{\sin P} = \frac{q}{\sin Q}$ becomes $\frac{QR}{\sin 27.7957725°} = \frac{13}{\sin 120°}$
   hence, $QR = \frac{13 \times \sin 27.7957725°}{\sin 120°} = 7$
   length of $QR = 7$ cm

4. a) $12 < 15$ so side $XZ <$ side YZ
   hence, angle Y < angle X and so angle Y < 40°

   b) $\frac{\sin X}{x} = \frac{\sin Y}{y} = \frac{\sin Z}{z}$ becomes
   $\frac{\sin 40°}{15} = \frac{\sin Y}{12} = \frac{\sin Z}{z}$
   hence, $\sin Y = \frac{12 \times \sin 40°}{15}$
   $= 0.514230087$
   angle Y is acute so $Y = \sin^{-1}(0.514230087)$
   $= 30.94600686°$
   angle $Y = 30.9°$ to 1 decimal place
   angle $Z = 180° - (40° + 30.94600686°)$
   $= 109.0539931°$
   angle $Z = 109.1°$ to 1 decimal place

   c) $\frac{x}{\sin X} = \frac{z}{\sin Z}$ becomes
   $\frac{15}{\sin 40°} = \frac{XY}{\sin 109.0539931°}$
   hence, $XY = \frac{15 \times \sin 109.0539931°}{\sin 40°}$
   $= 22.05731737$
   length of $XY = 22.1$ cm to 3 significant figures

EXERCISE 27

1. Using $b^2 = c^2 + a^2 - (2ca \cos B)$,
   $(AC)^2 = 10^2 + 12^2 - (2 \times 10 \times 12 \times \cos 45°)$
   $= 100 + 144 - (169.7056275)$
   $= 74.294337252$
   $AC = \sqrt{74.29437252}\,\text{cm} = 8.61941834\,\text{cm}$
   Length of AC $= 8.62$ cm to 3 significant figures.

2. Using $f^2 = d^2 + e^2 - (2de \cos F)$,
   $(DE)^2 = 9^2 + 14^2 - (2 \times 9 \times 14 \times \cos 150°)$
   $= 81 + 196 - (-218.2384018)$
   $= 81 + 196 + 218.2384018$
   $= 495.2384018$
   $DE = \sqrt{495.2384018}\,\text{cm} = 22.2539525\,\text{m}$
   Length of DE $= 22.3$ m to 3 significant figures.

3. Use $p^2 = q^2 + r^2 - (2qr \cos P)$ or
   $\cos P = \frac{q^2 + r^2 - p^2}{2qr}$
   That is $9^2 = 8^2 + 11^2 - (2 \times 8 \times 11 \times \cos P)$ or
   $\cos P = \frac{8^2 + 11^2 - 9^2}{2 \times 8 \times 11}$
   These give $\cos P = \frac{104}{176} = 0.59090909$
   $P = \cos^{-1}(0.59090909) = 53.77845338°$
   Angle P $= 53.8°$ to 1 decimal place.

4. a) Using $s^2 = t^2 + u^2 - (2tu \cos S)$,
   $(TU)^2 = 15^2 + 10^2 - (2 \times 15 \times 10 \times \cos 95°)$
   $= 225 + 100 - (-26.14672282)$
   $= 225 + 100 + 26.14672282$
   $= 351.1467228$
   $TU = \sqrt{351.1467228}\,\text{m} = 18.73890933\,\text{m}$
   Length of TU $= 18.7$ m to 3 significant figures.

   b) $\frac{\sin S}{S} = \frac{\sin T}{t} = \frac{\sin U}{u}$ becomes
   $\frac{\sin 95°}{18.73890933} = \frac{\sin T}{15} = \frac{\sin U}{10}$
   Hence, $\sin U = \frac{10 \times \sin 95°}{18.73890933}$
   $= 0.531618292$
   Angle U is acute so $U = \sin^{-1}(0.531618292)$
   $= 32.11486168°$
   Angle U $= 32.1°$ to 1 decimal place.

   c) Angle T $= 180° - (95° + 32.11486168°)$
   $= 52.88513832°$
   Angle T $= 52.9°$ to 1 decimal place.
   Alternatively,
   $\frac{\sin T}{15} = \frac{\sin 95°}{18.73890933}$ gives $\sin T = \frac{15 \times \sin 95°}{18.73890933}$
   $= 0.797427438$
   Hence, $T = \sin^{-1}(0.797427438)$
   $= 52.88513832°$
   Angle T $= 52.9°$ to 1 decimal place.

5. a) $\cos X = \frac{y^2 + z^2 - x^2}{2yz} = \frac{8^2 + 15^2 - 13^2}{2 \times 8 \times 15}$
   $= \frac{64 + 225 - 169}{240} = \frac{120}{240} = 0.5$
   Hence, $X = \cos^{-1}(0.5)$ and so angle X $= 60°$

   b) $\cos Y = \frac{z^2 + x^2 - y^2}{2zx} = \frac{15^2 + 13^2 - 8^2}{2 \times 15 \times 13}$
   $= \frac{225 + 169 - 64}{390} = \frac{330}{390} = \frac{11}{13}$
   Hence, $Y = \cos^{-1}(0.846153846)$
   $= 32.2042275°$ and so
   angle Y $= 32.2°$ to 1 decimal place.
   Alternatively, the sine rule
   $\frac{\sin Y}{y} = \frac{\sin X}{x}$ gives $\frac{\sin Y}{8} = \frac{\sin 60°}{13}$
   Hence, $\sin Y = \frac{8 \times \sin 60°}{13} = 0.53293871$
   Y is opposite the shortest side of the triangle so it is the smallest angle. It follows that Y is acute and $Y = \sin^{-1}(0.53293871) = 32.2042275°$.
   Angle Y $= 32.2°$ to 1 decimal place.

   c) $\cos Z = \frac{x^2 + y^2 - z^2}{2xy} = \frac{13^2 + 8^2 - 15^2}{2 \times 13 \times 8}$
   $= \frac{169 + 64 - 225}{208} = \frac{8}{208} = \frac{1}{26}$
   $Z = \cos^{-1}(0.038461538) = 87.7957725°$ and so angle Z $= 87.8°$ to 1 decimal place.
   Alternatively, $Z = 180° - (60° + 32.2042275°)$
   $= 87.7957725°$
   Angle Z $= 87.8°$ to 1 decimal place.

EXERCISE 28

1. The table of values for the function $y = \cos x°$ is shown below.

x	0	30	60	90	120	150	180	210	240	270	300	330	360
y	1	0.87	0.5	0	−0.5	−0.87	−1	−0.87	−0.5	0	0.5	0.87	1

   Here is the graph:

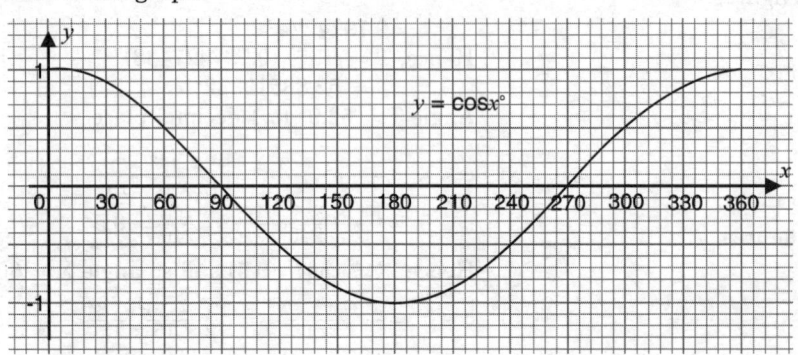

   [Graph of $y = \cos x°$]

2. a) The missing entries are shown here:

x	60	105	150	195	240	285	330
y	0.37	1.22	1.37	0.71	−0.37	−1.22	−1.37

   Note: When you have completed the table, you should notice patterns in the sequence of y-values. This helps in detecting errors.

   b) Here is the graph:

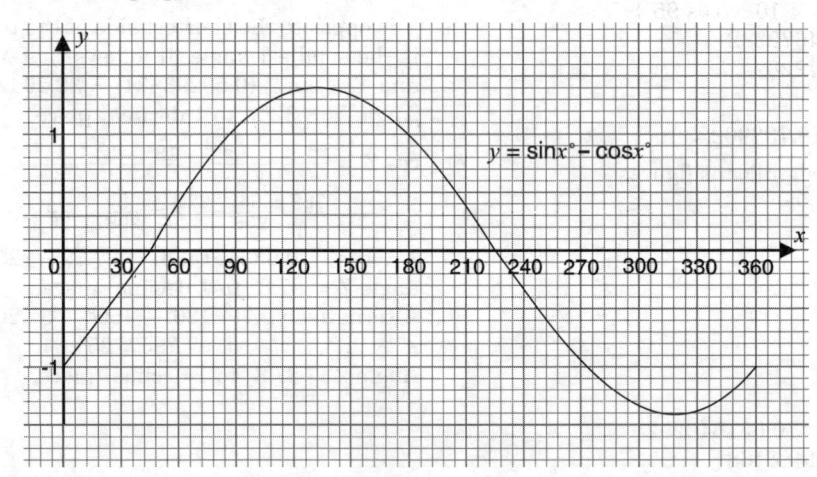

   [Graph of $y = \sin x° - \cos x°$]

3. a) The missing entries are shown here:

x	30	120	210	330
y	2	2.73	0	0

   b) Here is the graph:

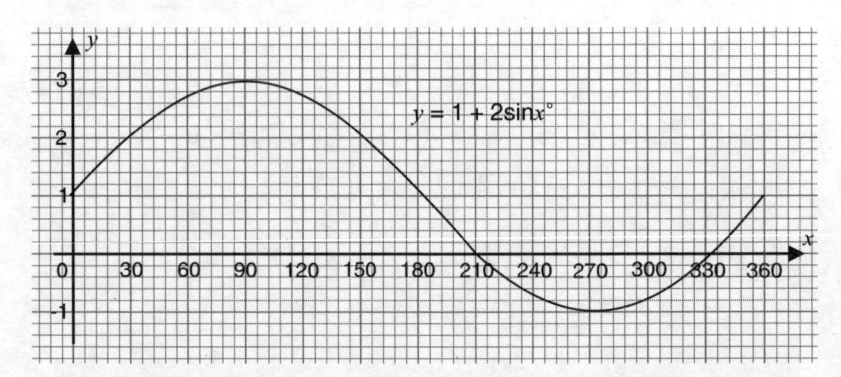

   [Graph of $y = 1 + 2\sin x°$]

EXERCISE 29

1. a) Triangle ABC has a right angle at B, so we apply Pythagoras's theorem.
   $(AC)^2 = (AB)^2 + (BC)^2 = 20^2 + 15^2$
   $= 400 + 225 = 625$
   $AC = \sqrt{625}$ cm
   Length of AC = 25 cm

   b) Triangle BCE has a right angle at C so
   $\tan 41° = \frac{\text{opp}(41°)}{\text{adj}(41°)} = \frac{EC}{15}$
   $EC = 15 \times \tan 41° = 13.03930107$ cm
   Length of EC = 13.0 cm to 3 significant figures.

   c) AC is the projection of AE on the plane ABCD. The angle required is EÂC. Triangle EAC has a right angle at C.
   $\tan(E\hat{A}C) = \frac{\text{opposite}}{\text{adjacent}} = \frac{EC}{AC}$
   $= \frac{13.03930107}{25} = 0.521572042$
   Angle $EAC = \tan^{-1}(0.521572042)$
   $= 27.54528604°$
   Angle which AE makes with the horizontal = 27.5° to 1 decimal place.

2. a) In triangle ABC, angle $A = 180° - 38° = 142°$
   (angles on a straight line)
   and angle $C = 38° - 22° = 16°$
   (using exterior angle of triangle property)

   The sine rule
   $\frac{a}{\sin A} = \frac{b}{\sin B} = \frac{c}{\sin C}$ gives
   $\frac{BC}{\sin 142°} = \frac{AC}{\sin 22°} = \frac{50}{\sin 16°}$
   Hence, $BC = \frac{50 \times \sin 142°}{\sin 16°} = 111.679615$ m
   Distance BC = 112 m to 3 significant figures.

   b) Triangle BCD has a right angle at D, so
   $\sin 22° = \frac{\text{opp}(22°)}{\text{hypotenuse}} = \frac{CD}{BC} = \frac{CD}{111.67915}$
   Hence, $CD = 111.67915 \times \sin 22°$
   $= 41.83592011$ m
   Height of the building = 41.8 m to 3 significant figures.

3. a) Using the cosine rule,
   $(OS)^2 = 400^2 + 850^2 - (2 \times 400 \times 850 \times \cos 110°)$
   $= 160\ 000 + 722\ 500 - (-232\ 573.6975)$
   $= 160\ 000 + 722\ 500 + 232\ 573.6975$
   $= 1\ 115\ 073.6975$
   $OS = \sqrt{1\ 115\ 073.6975}$ km = 1055.9705 km
   Distance from Osaka to Sapporo = 1060 km to 3 significant figures.

   b) Using the sine rule in triangle OST
   $\frac{\sin(O\hat{S}T)}{850} = \frac{\sin 110°}{1055.9705}$
   $\sin(O\hat{S}T) = \frac{850 \times \sin 110°}{1055.9705} = 0.756402501$

   Angle OST is acute so $O\hat{S}T = \sin^{-1}(0.756402501)$
   $= 49.14806859°$
   Angle OST = 49° to the nearest degree.

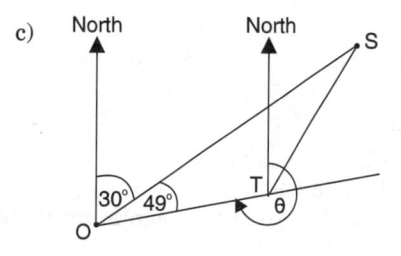

   c) Bearing of Tokyo from Osaka
   $= 079° (30° + 49°)$
   Bearing of Osaka from Tokyo
   $= \theta$ in the diagram
   $= 79° + 180°$
   Bearing of Osaka from Tokyo = 259°.

   d) Time taken $= \frac{\text{distance}}{\text{speed}} = \frac{850 \text{ km}}{500 \text{ km/h}} = 1.7$ hours
   $= 1$ hour 42 minutes.
   Time of arrival in Tokyo = 1112

4. a) (i) Depth of water at 2 a.m.
   $= (4 - 3 \sin 60°)$ m
   $= 1.401923789$ m
   $= 1.40$ m to 3 significant figures
   (ii) Depth of water at 10 a.m.
   $= (4 - 3 \sin 300°)$
   $= 6.598076211$ m
   $= 6.60$ m to 3 significant figures

   b) The missing entries are shown here:

t	2	4	7	10
d	1.4	1.4	5.5	6.6

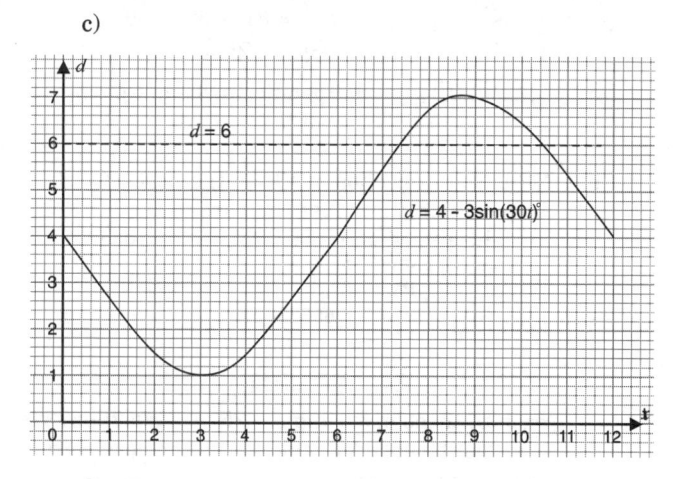

   c)

   d) Between 7.18 a.m. and 10.30 a.m.

   notice that each small square on the t-axis represents 12 minutes

Check your progress 4

1. a) Using the cosine rule,
   $AB^2 = 8^2 + 3^2 - (2 \times 8 \times 3 \times \cos 15°)$
   $= 64 + 9 - (46.36443966)$
   $= 26.63556034$
   $AB = \sqrt{26.63556034}$ m
   $= 5.160965059$ m
   Length of AB $= 5.16$ m to 3 significant figures.

   b) Area of triangle OAB $= \frac{1}{2} \times 8 \times 3 \times \sin 15°$ m^2
   $= 3.105828541$ m^2
   Area of triangle OAB $= 3.11$ m^2 to 3 significant figures.

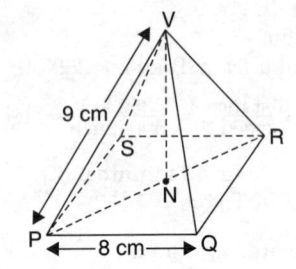

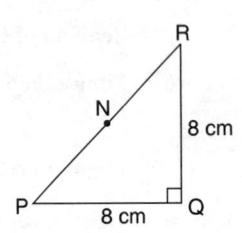

2. a) VN is the perpendicular from V to the base PQRS. By symmetry, N is the centre of the square PQRS.
   Triangle PQR has a right angle at Q, so by Pythagoras's theorem
   $(PR)^2 = 8^2 + 8^2 = 64 + 64 = 128$
   $PR = \sqrt{128}$ cm $= 11.3137085$ cm
   $PN = \frac{PR}{2} = 5.656854249$ cm

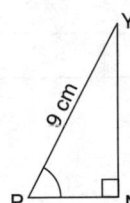

   Triangle VPN has a right-angle at N, so by Pythagoras's theorem,
   $9^2 = (VN)^2 + (5.656854249)^2$
   $(VN)^2 = 81 - 32 = 49$
   $VN = \sqrt{49} = 7$ cm
   Perpendicular height of the pyramid $= 7$ cm.

   b) NP is the projection of VP on the base PQRS. Angle which VP makes the base = angle VPN.
   In right-angled triangle VPN, we can use
   $\sin(V\hat{P}N) = \frac{VN}{PV} = \frac{7}{9}$ or
   $\cos(V\hat{P}N) = \frac{PN}{PV} = \frac{5.656854249}{9}$ or
   $\tan(V\hat{P}N) = \frac{VN}{PN} = \frac{7}{5.65854249}$
   Each of these gives angle VPN $= 51.05755873°$ so angle the edge VP makes with the base $= 51.1°$ to 1 decimal place.

3. a) y-coordinate of A $= \sin 90° = 1$ so coordinates of A are (90,1).
   b) $\sin 270° = -1$.
   c)

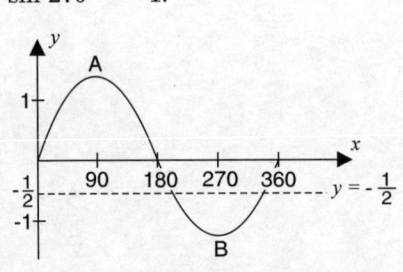

   d) The two graphs intersect at two points so, for $0 \le x \le 360$, the equation $\sin x = -\frac{1}{2}$ has two solutions.

4. a) (i) $p^2 = a^2 + b^2 - (2\,ab\,\cos P)$
   $(AB)^2 = 100^2 + 60^2 - (2 \times 100 \times 60 \times \cos 80°)$
   $= 10\,000 + 3600 - (2083.778132)$
   $= 11\,516.22187$
   $AB = \sqrt{11\,516.22187}$ km
   $= 107.3136611$ km
   $AB = 107$ km to 3 siginificant figures.

   (ii) Using the sine rule in triangle PAB,
   $\frac{\sin P}{p} = \frac{\sin A}{a} = \frac{\sin B}{b}$ becomes
   $\frac{\sin 80°}{107.3136611} = \frac{\sin A}{100} = \frac{\sin B}{60}$
   Hence, $\sin A = \frac{100 \times \sin 80°}{107.3136611}$
   $= 0.917690947$
   AB is the longest side of the triangle so P is the largest angle. Hence, angle A is acute.
   $A = \sin^{-1}(0.917690947) = 66.59081488°$
   Angle PAB $= 66.6°$ to 1 decimal place.

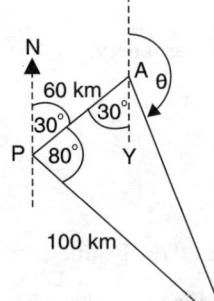

   (iii) Bearing of B from A $= \theta$ in the diagram.
   Angle YAB $= 66.59081488° - 30°$
   $= 36.59081488°$
   $\theta = 180° -$ angle YAB
   $= 143.4091851°$
   Bearing of B from A $= 143°$ to the nearest degree.

   b) (i) The faster ship travels the greater distance, that is 100 km.
   Time $= \frac{\text{distance}}{\text{speed}}$ so
   $t = \frac{100}{20}$
   that is $t = 5$

   (ii) Speed of slower ship $= \frac{\text{distance}}{\text{time}}$
   $= \frac{60 \text{ km}}{5 \text{ hours}}$
   $= 12$ km/h

Index

A
area 6
    of a circle 21
    of a parallelogram 12
    of a rectangle 8
    of a sector 38
    of a trapezium 12
    of a triangle 11, 127
    units of 9
area scale factor 32
area, surface (*see* surface area)

C
capacity 60
centimetre 2
circle
    area of a 21
    circumference of a 16
    segment of a 40
circular arc
    length of a 37
circumference
    of a circle 16
cone
    surface area of a 69
    volume of a 69
cos(angle) 107
cosine
    function 140
    ratio 106
    rule 134
cuboid
    surface area of a 46
    volume of a 52
cylinder
    surface area of a 49
    volume of a 57

D
density 64

E
easting 126

F
function
    cosine 140
    sine 140

G
gram 64

H
height
    perpendicular 11, 69
    slant 69
hemisphere 70
hypotenuse 82

K
kilogram 64
kilometre 2

L
length 1
    of a circular arc 37
linear scale factor 32, 73
litre 60

M
mass 64
metre 1
    square 7
millimetre 2

N
northing 126

P
parallelogram
    area of a 12
perimeter 3
perpendicular height 11, 69
pi 17
polygon 45
polyhedron 45
prism
    volume of a 55
pyramid
    surface area of a 68
    volume of a 68
Pythagoras's theorem 81

Q
quadrant 38

R
ratio(s)
    cosine 106
    sine 106
    tangent 96
rectangle
    area of a 8
rectangular box 48
rule
    cosine 134
    sine 131

S
scale factor
    area 32
    linear 32, 73
    surface area 73
sector
    area of a 38
    of a circle 38
segment
    of a circle 40
semicircle 38
sin(angle) 107
sine
    function 140
    ratio 106
    rule 131
slant height 69
sphere
    surface area of a 70
    volume of a 70
square metre 7
solid
    surface area of a 45
surface area
    of a cone 69
    of a cuboid 46
    of a cylinder 49
    of a pyramid 68
    of a solid 45
    of a sphere 70
    scale factor 73

T
tan(angle) 97
tangent ratio 96
tonne 64
trapezium
    area of a 12
triangle
    area of a 11, 127
    measurement 81
trigonometry 81

V
volume 51
    of a cone 69
    of a cuboid 52
    of a cylinder 57
        of a prism 55
    of a pyramid 68
    of a sphere 70
    units of 52

W
weight 64